Rethinking Global Urbanism

Routledge Advances in Geography

1 **The Other Global City**
Edited by Shail Mayaram

2 **Branding Cities**
Cosmopolitanism, Parochialism, and Social Change
Edited by Stephanie Hemelryk Donald, Eleonore Kofman and Catherine Kevin

3 **Transforming Urban Waterfronts**
Fixity and Flow
Edited by Gene Desfor, Jennefer Laidley, Dirk Schubert and Quentin Stevens

4 **Spatial Regulation in New York City**
From Urban Renewal to Zero Tolerance
Themis Chronopoulos

5 **Selling Ethnic Neighborhoods**
The Rise of Neighborhoods as Places of Leisure and Consumption
Edited by Volkan Aytar and Jan Rath

6 **The Gentrification of Nightlife and the Right to the City**
Regulating Spaces of Social Dancing in New York
Laam Hae

7 **Rethinking Global Urbanism**
Comparative Insights from Secondary Cities
Edited by Xiangming Chen and Ahmed Kanna

Rethinking Global Urbanism

Comparative Insights from Secondary Cities

Edited by Xiangming Chen and Ahmed Kanna

NEW YORK LONDON

First published 2012
by Routledge
711 Third Avenue, New York, NY 10017

Simultaneously published in the UK
by Routledge
2 Park Square, Milton Park, Abingdon, Oxon OX14 4RN

Transferred to digital print 2013

Routledge is an imprint of the Taylor & Francis Group, an informa business

First issued in paperback 2013

Library of Congress Cataloging-in-Publication Data

Rethinking global urbanism : comparative insights from secondary cities / edited by Xiangming Chen and Ahmed Kanna.
p. cm. — (Routledge advances in geography ; 7)
Includes bibliographical references and index.
1. Cities and towns. 2. Urbanization. 3. Sociology, Urban. 4. Globalization—Social aspects. I. Chen, Xiangming, 1955– II. Kanna, Ahmed.
HT119.R48 2012
307.76—dc23
2011047889

ISBN13: 978-0-415-89223-0 (hbk)
ISBN13: 978-0-203-11836-8 (ebk)
ISBN13: 978-0-415-72030-4 (pbk)

Typeset in Sabon
by IBT Global.

Contents

Maps

Figures

Tables

Acknowledgments

Xiangming Chen: This book is the first major publication developed and sponsored by the Center for Urban and Global Studies (CUGS) at Trinity College in Hartford, Connecticut. The genesis of the book was a conference entitled "Rethinking Cities and Communities: Urban Transition before and During the Era of Globalization," which took place on Trinity campus on November 14–15, 2008. By organizing this conference, the fledging CUGS, which was launched on October 19, 2007, began to put itself on the intellectual map of global urban research through scholarly dialogues with urban scholars everywhere. The conference provided an excellent forum for a number of Trinity faculty to engage in discussions and debates with a number of senior and junior scholars of cities and communities from Harvard, MIT, Yale, Brown, and other research universities and liberal-arts colleges in New England and beyond. From all the papers presented at the conference, a number was selected to be revised for publication in this book, together with a couple of newly invited contributions including the preface written by Tim Bunnell and James Sidaway. The book has ended up with an even balance of authors from other academic institutions and Trinity College or those who were recently affiliated with Trinity as visiting scholars, including my coeditor Ahmed Kanna, who was the inaugural Paul Raether Postdoctoral Fellow in Urban Studies at CUGS during 2008–09. It was great to have him as a colleague during 2008–09 and to have co-edited this book with him since then.

First of all, I would like to express my deep gratitude to the Andrew Mellon Foundation for a major grant and the generous financial support from donors to the Trinity College Mellon Challenge for Urban and Global Studies that endowed CUGS and have supported its vibrant growth and accomplishments as exemplified by the publication of this book. Second, I owe thanks to a variety of people at Trinity College, including Paul E. Raether, chairman of the Board of Trustees, all the Trustees, President James F. Jones, Jr., Dean of Faculty Rena Fraden, many faculty colleagues and friends, dedicated staff working at CUGS, and a growing number of students who have been contributors to CUGS as research assistants and in other capacities. Three Trinity students and recent graduates, Tomas de'Medici '11, Chang Liu '12, and Michael Magdelinskas '11, became so involved in this project that they ended up as coauthors, whereas recent Trinity graduates Nick Bacon '10, Tomas de'Medici '11, Henry Fitts '12, Brooke Grasberger '12, Michael Magdelinskas '11, Curtis Stone '10, and Yuwei Xie '11 lent their hands in commenting on and editing some of these chapters. I would like to single out Teresita Romero for her superb assistance in the final stage of editing, formatting, and indexing the book, while Byron H. Chen contributed to proofreading. We also thank Eleanor Chan and Michael Watters of Integrated Book Technology, Inc for overseeing the production of the book. Thanks go to Max Novick and Jennifer Morrow of Routledge for their trust and support and to the anonymous reviewers for their comments and criticisms that have helped improve the book. I am grateful to Dennis Judd, Jennifer Robinson, and Saskia Sassen for their kind words that grace the back of this book. Finally, I would like to thank all the contributors for their support, cooperation, and patience, without which this book would not have been possible.

Xiangming Chen
Hartford, Connecticut
October 2011

Ahmed Kanna: I would like to thank Xiangming Chen and Vijay Prashad for inviting me to apply for the Raether Postdoctoral Fellowship at Trinity in 2008–2009. Xiangming, Vijay, my other Trinity Colleagues, and my Trinity students were all wonderful interlocutors during the year of my research fellowship. The contributors to this volume were generous with their effort, time, and deep scholarly knowledge, and have made the final product a valuable contribution to the global urbanism literature. I thank, as well, Max Novick and Jennifer Morrow at Routledge, whose guidance and patience through the editorial process has been invaluable. Tim Bunnell and James Sidaway, who wrote the preface, and Gary McDonogh, who generously agreed to blurb the volume, all deserve my great thanks.

Ahmed Kanna
Oakland, California
October 2011

Preface

Tim Bunnell and James D. Sidaway

As Ahmed Kanna and Xiangming Chen's introduction to *Rethinking Global Urbanism* sets out in more detail, it is by now widely noted that urban scholars need to expand the range of cities which they examine in relation to processes of globalization beyond the three archetypes of leading 'global cities' (London, New York, Tokyo) famously proposed by Saskia Sassen (1991). Indeed, soon afterwards Manuel Castells argued that 'the global city phenomenon cannot be reduced to a few urban cores at the top of the hierarchy' (Castells 1996: 380). Subsequently, Sassen (2002: 3) herself has been among those to bring cities of 'the Global South and in the mid-range of the global hierarchy' into the literature. Although there were other precursors, such as Simon (1992), it is work marshaled by the online Globalization and World Cities (GaWC) research network that has since helped to extend coverage (http://www.lboro.ac.uk/gawc/). But this too is supplemented by those who focus on non-Western cities as sites of cultural exchange and production (Huyssen 2008).

What has not changed much, however, is the tendency for scholars to order the cities that they examine in a rather hierarchical fashion or to approach them through reference to the sociospatial structure and architecture of selected Western cities (originally perhaps Paris, Berlin and London, then New York and Chicago and latterly Los Angeles; see Nicholls 2010). Whereas Sassen's 'mid-range' was determined in relation to presumably higher and lower-range cities, researchers connected with GaWC began by identifying 'alpha,' 'beta' and 'gamma' world cities (Beaverstock et al. 1999) on grounds of their levels of connectivity and concentration of command and control functions. *Rethinking Global Urbanism* is therefore especially welcomed, both because of the range of cities that it incorporates and also because it resists the ingrained tendency to examine such diversity hierarchically. Moreover, the book not only includes chapters that treat these less-studied secondary cities holistically (city as content or study *of* the city) but also features contributions that approach certain urban phenomenon in these cities (city as context or study *in* the city).

The chapters assembled by Xiangming Chen and Ahmed Kanna build on such critiques of the hierarchical ranking tendencies of the global/

world cities literature. Moreover, the volume pushes beyond hierarchical framings in ways that add empirical substance to calls for doing urban studies differently. Another key voice here is Jennifer Robinson, whose *Ordinary Cities: Between Modernity and Development* (also published by Routledge) explored agendas for 'postcolonialising' urban studies (Robinson 2006: 1). This calls for approaches resisting the ingrained tendency for urban theory to take its primary inspiration from cities in the West. While the world/global cities literature has expanded to incorporate some cities that might, until recently, have been classified as "Third World," the continued assumption of hierarchical relations continues to present alpha (or, most recently, "alpha ++") cities as the leading edge of urban innovation, dynamism and aspiration. For Robinson, seeing *all* cities as "ordinary" thus means "gathering difference as diversity rather than as hierarchical division" (p. 6). Yet putting this into practice is difficult. The assemblage of cities and urban regions including Bilbao, Detroit-Windsor, Dompak, Dubai, Harare, Johannesburg, Kunming, Putrajaya, Salvador, San Francisco, Shenzhen, Singapore, Springfield, Tianjin and Tunis in this volume makes *Rethinking Global Urbanism* a marker on the way to a more global and open urban studies.

Rethinking Global Urbanism also marks a coming of age within the recent proliferation of comparative urban studies. This includes a strong strand of postcolonially inflected scholarship (McFarlane 2010; Robinson 2011), and allied material appearing in the *International Journal of Urban and Regional Research*. But it is also intertwined with wider debates in urban geography, urban planning and cognate fields (see Nijman 2007 and other contributions to the volume that he introduces; and Ward 2010). The clear challenge in assembling diverse cases is that much of the explicitly comparative labor is left to readers traversing the chapters. *Rethinking Global Urbanism* undoubtedly reveals and will foment new comparative insights in diverse readers. This enables the logical next step, to develop projects that take such comparative insight and associated questions as their starting points.

How can questions arising from nonhierarchical and contextualized approaches to a world of cities form the basis of collective, relational comparative work? This is not a question that we claim to be able to answer definitively. However, it seems to us that collaborative methodological strategies suggest one possible way forward. Bringing diverse and distant cities into comparative conversation with each other requires insights and skills—not least proficiency in multiple languages other than English—that are unlikely to be found in a single scholar or perhaps even in any one institution. The regional partitioning of knowledge and associated academic training is, of course, at least in part a legacy of area studies. However, we do not see area-studies training as the problem but rather as an important resource for new collaborative urban possibilities. Nonetheless, as many of the authors whose work appears in the pages that follow will surely testify, this demands going beyond traditional, bounded approaches to area-based knowledge (Chen 2010).

Studies of global urban dynamics and linkages without nuanced immersion in cultures and languages other than English inevitably skew research to those in many ways highly unrepresentative urban worlds of Anglophone elites. One outcome of this is a paradoxical parochialism of some ostensibly 'global' research (Bunnell and Maringanti 2010). In encouraging contrast, the scholarship assembled in this volume engages with the global dimensions of the cities in ways which extend beyond the elite, top-down and advanced service and financial sector networks that have predominated in the world cities literature. As such, we believe that *Rethinking Global Urbanism* will inspire others to participate in the exciting challenge of remapping the contours of what it means to be urban in the twenty-first century.

REFERENCES

Beaverstock, Jonathan V., Richard G. Smith and Peter J. Taylor (1999). "A Roster of World Cities." *Cities* 16: 445–458.

Bunnell, Tim, and Anant Maringanti (2010). "Practicing Urban and Regional Research Beyond Metrocentricity." *International Journal of Urban and Regional Research* 34 (2): 415–420.

Castells, Manual (1996). *The Rise of the Network Society*. Cambridge, MA: Blackwell.

Chen, Kuan-Hsing (2010). *Asia as Method: Towards Deimperialization*. Durham, NC, and London: Duke University Press.

Huyssen, Andreas (2008). *Other Cities, Other Worlds: Urban Imaginaries in a Globalizing Age*. Durham, NC, and London: Duke University Press.

McFarlane, Colin (2010). "The Comparative City: Knowledge, Learning, Urbanism." *International Journal of Urban and Regional Research* 34 (4): 725–742.

Nicholls, Walter J. (2010). "The Los Angeles School: Difference, Politics, City." *International Journal of Urban and Regional Research* 35 (1): 189–206.

Nijman, Jan (2007). "Introduction—Comparative Urbanism." *Urban Geography* 28: 1–6.

Robinson, Jennifer (2006). *Ordinary Cities: Between Modernity and Development*. Abingdon, UK: Routledge.

Robinson, Jennifer (2011). "Cities in a World of Cities." *International Journal of Urban and Regional Research* 35 (1): 1-23.

Sassen, Saskia (1991). *The Global City: New York, London, Tokyo*. Princeton, NJ: Princeton University Press.

Sassen, Saskia, ed. (2002). *Global Networks, Linked Cities*. London: Routledge.

Simon, David (1992). *Cities, Capital and Development: African Cities in the World Economy*. London and New York: Belhaven Press and Halsted Press.

Ward, Kevin (2010). "Towards a Relational Comparative Approach to the Study of Cities." *Progress in Human Geography* 34 (4): 471–487.

this demands going beyond traditional, bounded approaches to area-based knowledge (Chen 2010).

Introduction

Bringing the Less Familiar Cities In and Together

Ahmed Kanna and Xiangming Chen

This book is a reflection on the theme of urban globalization in which we bring together a collection of essays on cities that are not usually part of conversations about globalization or global cities (see Map I.1). This will at first appear to be a counterintuitive approach. The literature on global cities has tended to focus on the global economy's seemingly dominant nodes: New York, London, Tokyo, Paris, Hong Kong, and more recently, Shanghai and Mumbai. Hardly any of the cities examined in this volume approach the physical scope, demographic size or economic power of these world metropolises. In some cases, the volume's contributors write about urban areas caught up in trajectories radically different from that of globalization's winners: cities and urban regions, in Brent Ryan's words (in Chapter 5), that inexorably spiral toward a condition of "de-globalization." Our unconventional choice of case studies, however, is directly connected to our argument in this book: "secondary," less well-examined cities bring better to light global processes that have been marginalized or neglected in the literature on global cities. Such processes include the emergence of alternative and new cartographies of globalization (Dawson and Edwards 2004: 2); the role of local, regional, and "deep" (economic, colonial, national) histories in shaping contemporary urban globalization; and the multifarious, complex role of cultural and symbolic structures in urban experience and the construction of global urban circuits. By focusing on this diverse set of "secondary" cities that are also "global" in different ways, we hope to chart a few new pathways or sideways through the somewhat familiar terrain of the global city scholarship.

In her pioneering book (1991), Saskia Sassen led the way of understanding how the systematic attributes of global capital require certain territorial moments that varied between New York, London, and Tokyo, and how the number and spatial forms of global cities increased in the 1990s and 2000s as corporate globalization has expanded (also see Sassen 2012). Much of the literature on global urbanism since Sassen's earlier work has been about "an integrating world economy, a homogenizing global culture and a coherent global polity" (Short 2006: 65). Future research on urban globalization, he suggests, should focus instead on "globalization as a process that generates fractured economies, splintering cultures and resurgent nation states" (Short

2006: 65).[1] Whereas we agree with Short's attempt to push urban studies away from a too monolithic framing of globalization, we go further in this volume. The collected essays highlight aspects of urban globalization that move more radically away from the most influential scholarly contributions (e.g., Friedman and Wolff 1982; Hall 1966; Sassen 1991), studies that focus on the planning and economic dimensions of urbanization (Short 2006: 66). Valuable as these studies have been, they have been underpinned by implicit (and sometimes explicit) notions of a global urban hierarchy (hence the focus on financial and command centers, which exemplify one kind, albeit a decisively powerful one, of globalization). Brent Ryan, in Chapter 5 of this volume, has come up with perhaps the most powerful formulation of our skepticism of the dominant literature. A certain kind of globality, he writes, summed up in the term "global city status" and reflecting a neoliberal, econo-centric orientation in urban policy and spatialization, peripheralizes questions of urban democracy and human rights. Moreover, "reinforcing the presumed desirability of globality is its seemingly teleological aspect," which, in turn, privileges those cities already in possession of this globality. This logic marginalizes the types of globality not conforming to econo-centric hierarchies and "mitigates against the substantial broadening of membership" in the privileged club of "global cities."

NEW CARTOGRAPHIES OF THE GLOBAL URBAN

In his theoretical survey, Short points out that globalization "act(s) in and through" all cities: "almost all cities can act as a gateway for the transmission of economic, political and cultural globalization" (Short 2006: 74). We agree, so far as this argument goes. But we caution against what still appear to be the vestiges of a reifying language which opposes "globalization" and "the city," such that the former, like an external object, acts upon or through the latter. This formulation implies both that urban areas are bounded entities and that globalization is a top-down, primarily economic process originating in the dominant nodes of the global economy (Short 2006: 74–75). As Sarah Moser, citing Arab cities scholar Yasser Elsheshtawy (2008), argues in Chapter 9: "The narrow economic focus of the global cities discourse excludes cities such as Jerusalem, Medina and Mecca which are surely global in the sense that they serve as religious centers that attract millions of international pilgrims annually. However, because they do not host headquarters for multinational corporations, they are not considered 'global' cities."

Among the contributions we seek to make to the discussion of global urbanism is to point to what such new, alternative cartographies of the global (Dawson and Edwards 2004) look like in the emerging order of the twenty-first century. Using their familiarity with a number of somewhat unfamiliar cities, the contributors collectively reveal multiple layers of the global urban landscape that have tended to be hidden and latent. We also seek to disrupt the assumption that globalization is a teleological narrative of capitalism's triumph, showing instead other powerful political and cultural currents that impinge on the economic and spatial unfolding of neoliberal logics.

Strongly evident in Zukin's opening chapter is such an appreciation of neoliberalism's constructedness and how an "aesthetic mode of production" services the remaking of cities as neoliberal spaces. One of the major consequences of the shift from the Keynesian to the neoliberal city, Zukin argues, has been the ascent in urban planning and design of an "artistic mode of production" (see also Zukin 1989). Museums, artists' enclaves, and striking buildings by superstar architects have become central to projects of urban entrepreneurialism and redevelopment (see also Broudehoux 2007; Kanna 2011). Major world cities such as New York and Shanghai become linked with aspiring, or "secondary," world cities (Bilbao, in Zukin's argument, but the point can be extended to other examples in the volume, such as Dubai, Detroit-Windsor, Kunming and even the somewhat idiosyncratic Putrajaya-Malaysia and Dompak-Indonesia). As the model of the industrial city anchoring a modernizing national economy recedes further into the past, a global circuit of ideas and symbolic capital emerges in which all cities strive toward a single ideal: the city of spectacularized urbanscapes geared towards consumption and tourism.

Tyanai Masiya (Chapter 8) gives specific African examples of such a process of localization. His study of participatory budgeting in Harare and Johannesburg examines the outcomes that emerge from the circulation of urban expertise (a parallel to the circulating images and types of expertise related to the city of consumer spaces, discussed by Zukin). Participatory budgeting, of which the Brazilians were pioneers in the late 1980s, has been more recently adapted in Africa, more successfully in Johannesburg, less so in Harare. For Masiya, such Global South intellectual-expertise exchanges are more relevant to and provide a more hopeful future for African urbanism than the modernist, state-centered projects of the postcolonial period (themselves examples of the global circulation of urban expertise; see Holston 1989). If the transnational transfer and imitation of cultural strategies and renewal projects end up homogenizing more cities (Zukin, this volume), the application of the concept and practice of participatory budgeting to unfavorable national political and local contexts can fail easily, as in Zimbabwe (Masiya, this volume).

In his study of the Chinese community of San Francisco in the first half of the twentieth century, Scott Tang (Chapter 3) gives us another concrete account of global-local adaptations, tensions, and "friction" (Tsing 2005) that produce cultural forms, political imaginaries, and social networks in urban contexts. In the nineteenth century, anti-immigration forces in San Francisco advanced their agenda by representing Chinatown as a den of urban iniquity. Eschewing a direct response to racism, Chinese San Franciscans used urban institutions—such as the business association known as the "Chinese Six Companies"—to represent them in interactions with non-Chinese political entities. Until the 1920s, the Chinese Six Companies retained a strong mainland Chinese cast: its leaders were recruited from China's examination-passing, gentry-scholar class.

Mainland politics, according to Tang, were also influenced by overseas communities such as the San Francisco Chinese. The Kuomintang (KMT), for example, depended heavily on overseas Chinese and actively recruited San Francisco Chinese during the 1920s. Its political fortunes were greatly enhanced by alliances it fashioned with the Chinese Six Companies and the Chinese War Relief Association. After 1949, the mainland KMT-Communist conflict spilled over into San Francisco. An ill-conceived terrorist campaign waged by the KMT against suspected leftists backfired, leading the vast majority of Chinatown residents to turn against Chiang and the KMT. Soon thereafter, mainland politics would again have an impact on Chinese Americans in San Francisco. The United States and the PRC's involvement in the Korean War would confront Chinese Americans with a difficult choice, with the tendency eventually being to stress assimilation to the United States over loyalties to the mainland. The case of San Francisco's Chinatown up to the period of the early Cold War shows that "the global does not simply shape the local or vice versa . . . Diasporic politics such as those in Chinatown have intertwined local and global roots and simultaneous local and global implications" (Tang, this volume).

Sarah Moser's study of Putrajaya-Malaysia and Dompak-Indonesia (Chapter 9) is another striking illustration of such less well-examined or emerging alternative cartographies of the global. Moser's analysis of these new administrative capitals demonstrates a process she terms "serial seduction," which questions the assumption that cities are stratified in a Western-dominated urban hierarchy. Both administrative capitals generally eschew an orientation to the West, opting instead for a model based on that of neighboring Singapore. For example, both of the new administrative capitals seek to emulate the synthesis of neoliberalism, public hygiene (or, less charitably, sterility), and greenery (in the form botanical gardens and elaborate parks) of their well-established Southeast Asian cousin. Moreover, both Putrajaya and Dompak actively participate in discourses of global Islam. Both cities attempt to embody a vision of the "progressive Muslim city" of the twenty-first century, a synthesis between neoliberal "best practices" and Muslim values. In turn, they are influencing other urban areas in the Muslim world: not only are "many little Putrajayas springing up in Malaysia and Indonesia, including Dompak," but cities such as Astana, Kazakhstan and Abuja, Nigeria, are following Putrajaya's lead, attempting to also become "hi-tech cities with Muslim values." According to Moser:

> This Islamic turn should be understood in the context of growing connections between the Malay world and the Middle East through 'Islamic' banking, development partnerships, and growing political ties . . . Putrajaya and Dompak are positioning themselves to attract the

> attention, and more importantly, the business of 'global' Muslims (i.e. the wealthy Arab elite) through tourism and investment.

Most importantly, Moser shows what one of the aforementioned alternative cartographies of globalization looks like. "Innovations and models of city development," she writes, "do not necessarily originate in the west."

New transnational maps, in a slightly different sense, have also been emerging more broadly across Asia during the past three decades. For example, the rise of the Tianjin Binhai New Area (TBNA), China, according to Chang Liu and Xiangming Chen (Chapter 7), represents a large global-city region that incorporated Sassen's insights about the denationalized, autonomous character of the global city (Sassen 1991). One of the fastest growing local economies in China, with average annual GDP growth of about 20 percent since 1996, TBNA has been dubbed "China's third growth engine" (after Shenzhen–Hong Kong and Pudong-Shanghai) by the Chinese media. However, "Tianjin is by no means a global city," write and Liu and Chen, "and may not even qualify for a globalizing city" due to its lack of financial service functions under the large shadow of nearby Beijing as a politically motivated powerful center of producer services.

> But TBNA can be regarded as a new node in the global economy . . . Hosting one of the busiest container ports in China and a busy international airport, TBNA is capable of serving as the gateway to a much larger city-region around Tianjin. Its favorable location in Northeast Asia attracts Korean and Japanese capital (OECD 2009). A logical follow-up question is, as the global city perspective suggests, whether and how global forces may make TBNA truly autonomous from its national political and economic anchors.

Drawing on Vogel (2009), Liu and Chen (this volume) argue that the global city-region can add both precision and richness to Sassen's theory, allowing an examination of urban regions that "simultaneously rise above and fall below the nation-state. The region around a global city signals a politically and economically more complex and extended space where cooperation may foster integration or where imbalanced development can brew tension."

While TBNA sheds light on a city-region embedded in a vast nation-state, the case of Dubai illuminates the interconnections between a city-state scale and a larger transnational region. In Chapter 2, Kanna shows how interconnected Dubai's fortunes are to the wider Eurasian and African regions in which it situated. Over the past forty years, Dubai has become an important node, if not one of the most important nodes, in a transnational region connecting the Middle East, the Indian Ocean, Africa, East Asia, and Russia. The factors at play—the rise of the UAE petro state, regional and world geopolitics—are many and complex. Among other things, the case of Dubai shows how a free-trade, politically semiautonomous city-state can act as an extraterritorial financial and export hub for other nation-states. In recent years, for example, Dubai has become, arguably, Iran's economic capital and the main reexport supplier for Iraq and Afghanistan (see Marchal 2005). Kanna's chapter (Chapter 2) further problematizes conventional notions of global urban hierarchy and the boundedness of global cities. If the notion of a hierarchical global urban network is sustainable, then—as Masiya, Tang, Moser, Chen and de'Medici, Liu and Chen, and Kanna all show—this framing must account for maps of the global which accommodate the globalization of non-Western types of symbolic capital and expertise, as well as new geographies of economic exchange and interconnection.

In considering the cases of two former motors of North American industrialism, Springfield, Massachusetts, and Detroit-Michigan–Windsor-Ontario, Robert Forrant (Chapter 4) and Brent Ryan (Chapter 5) both make the intriguing case that, although the capitals of global financial and services economy are still mainly located in the Global North, struggling and even dying cities are also becoming common. This is a condition that Ryan terms "de-globalization," by which he means that these cities are no longer the economic engines that had benefited from the expanding global economy and market. The results for these former industrial capitals have been catastrophic, in terms of job losses, urban decay, municipal budgetary crisis, and the death of the rich, complex social orders built upon these urban-centered economies. The solutions proposed by influential urban actors are predictably corporatist and neoliberal: the creation of consumption-oriented, "parasitic" megaprojects (as Ryan puts it) such as casinos, museums, and outdoor urban streets-cum-shopping malls. It is tempting to conclude that cities such as Springfield, Massachusetts, Detroit, Michigan, and Windsor, Ontario, are now disconnected from the main action of globalization. But as both Forrant and Ryan show, these city-regions' current struggles are a direct result of the shifts in the global economy now so well established in the literature. Moreover, their turn toward consumption and tourism as panaceas (e.g., Detroit's desperate but failed attempt to revive itself through casinos) reflects the phenomena described by Zukin in the first chapter: the globalization of neoliberal urban redevelopment strategies that look similar from Shanghai to Bilbao to the supposedly "de-globalizing" former industrial capitals of the North.

DEEP URBAN ECONOMIES AND HISTORIES

In a recent article entitled "Cities in Today's Global Age," Saskia Sassen advises urban scholars to be sensitive to the distinctive features of cities and urban regions. "The deep economic history of a place matters" and is important in shaping how a city develops. Homogeneity, according to Sassen, is not desirable for globalizing institutions because "the global economy needs diverse specialized economic capabilities" (Sassen 2009:

7–8). Globalization, Sassen suggests, is not homogeneous but multifarious, fractured, and plural. Different globalized circuits, moreover, connect particular countries and cities: Mumbai to London and Bogotá via real-estate investment; New York to Brazil, Kenya, and Indonesia via circuits of trade in coffee futures; London, New York, Chicago, and Zurich through gold as a financial instrument; São Paulo, Johannesburg, and Sydney through metal wholesale circuits; and Mumbai and Dubai via retail circuits (Sassen 2009: 5–6).

Furthermore, as both Forrant and Ryan demonstrate, this point can be extended to urban areas which have been adversely affected by global processes. The deep economic history of regions such as Detroit-Windsor and the Connecticut River Valley has profoundly shaped the ways in which these regions have adapted to major shifts in the global economy over the past half century. This often results in disconnection from the most profitable circuits of the global economy, but can bring with it other kinds of global interconnection, as can be seen, for example, in increasingly transnational, migrant demographics of deindustrializing cities such as Hartford and Springfield in the Connecticut River Valley. Writing about a different context, Liu and Chen show how the national state with its historically deep traditions of top-down centralization is involved in shaping China's economic trajectory and structuring China's urban articulations with the global economy (Chen 2009). Urbanization and city-building in this context are central to the position that China wants to assume as a leader in the twenty-first-century manufacturing and transshipment economies, with SEZs such as Shenzhen (see Chen and de'Medici, this volume), Pudong, and now TBNA spearheading this process.

Moving from strict economics to the political-economic, chapters in this volume engage the role of colonialism and empire in the development of global cities (see also Jacobs 1996; Kusno 2000; Srinivas 2001). Kanna (Chapter 2), for example, focuses on how the imperial encounter shaped both Dubai's and Singapore's postcolonial and global development. The tradition of being a politically authoritarian, economically free market entrepôt is a continuous theme through both cities' deep economic histories. Between the early nineteenth and mid-twentieth century, Singapore became central to British expansion in the Malay Peninsula, a focal point of free trade in Southeast Asia, and a premier global free port; Dubai was far more isolated and impoverished (again, as a result of British colonial policy), but it too was operating as a free port as early as the first decade of the twentieth century. Moreover, the governmental institutions in Dubai, as in other parts of the British-dominated Arab Gulf, still follow the mold crafted by the British in collaboration with local princes during the colonial period. Whereas relatively stable colonial structures remained in place in Dubai, in Singapore, the Japanese occupation (1942–1945) had an ironic effect: it both mobilized an antioccupation movement and fatally delegitimized British rule in the Malay region, setting the scene for the rise of the

People's Action Party (PAP) with, again, profound consequences for contemporary Singapore globalization and urbanism.

In Chapter 9, Moser also sheds light on how the colonial legacy continues to shape urban policy in Southeast Asia. In both Malaysia and Indonesia, Chinese residents were the majority of the urban populations when these countries gained independence (Indonesia in 1945, Malaysia in 1965). Colonial racial categorizations, according to Moser, influenced land occupations and urban settlement patterns.

> Various late-colonial policies encouraged Malays, who were believed to be inherently 'lazy' and ill-suited to mercantile activities, to live in rural areas. Because Chinese were believed to be inherently more advanced in business, in the 1800's Chinese nationals were encouraged by colonial administrations to pursue mercantile activities in urban regions, a pattern that persists to the present day.

The resulting Chinese control of the urban economy in Malaysia and the Riau Islands in Indonesia led to the introduction of policies and incentives to encourage Malays to settle in urban areas. (Ironically, these policies had the unintentional result of further ethnic segregation in urban areas.) "It is against this backdrop of racial tension and geographical divisions" in older cities "that new 'Malay' cities of Putrajaya and Dompak were conceived" (Moser, this volume)

Another aspect of the colonial legacy, according to Moser, is the normalization of urban-centric development in the Malay world. The identification between success and being urban is "deep-seated," entailing an association between the rural and backwardness. In undertaking the Dompak and Putrajaya projects, the Indonesian and Malaysian states adapt this urban-rural distinction to justify massive spending and ostentatious megaprojects.

CULTURE MATTERS IN MULTIPLE WAYS

Anthropologists and other students of culture generally look at cultural processes through one (or a combination) of at least three lenses: culture as a symbolic resource in political or economic agendas, culture as structure for experience, and culture as source of agency (i.e., cultural practices). Whereas the chapters by Zukin, and to some extent Forrant and Ryan, emphasize the first framing, discussing how, among other things, the urban is commodified and culture, in a narrow sense—artists' enclaves, museums, and zones of "bourgeois gratification" (Ghirardo 1991)—becomes a strong orientation of recent urban redevelopment, the chapters by Notar, Collins, and Perry focus on the latter two framings.

The theme of cultural experiences and practices as carriers and shapers of globalization has been underemphasized in global-cities literature (Dawson and Edwards 2004: 3–4; Short 2006: 65). The chapters by Notar, Collins, and Perry speak to this connection between culture and globalization. Drawing on David Harvey's notions of geographical imagination and spatial consciousness, Notar (Chapter 10) is concerned with how culturally shaped worldviews intersect with status and class in a rapidly globalizing provincial city in China. "China," writes Notar, "is rebuilding cities on an arguably unprecedented scale: there has been no other country which has engaged in such radical 'creative destruction' of *all* of its cities in such a short span of time." Kunming, caught up in this process and situated in the Greater Mekong Subregion, China's hub and pathway to Southeast Asia, is becoming an "intensely global" city. How do everyday people, especially nonelites, imagine new cityscapes and their fellow urban residents in such periods of change, uncertainty, and "liminality"? Notar focuses in particular on the experiences of Kunming taxi drivers, whose ranks consist of people who have not benefited from China's reform period: people who lost land or businesses, for example, as an indirect result of the reforms. This provides Notar a good vantage point from which to illuminate the intersection of cultural structures of meaning and the experience of urban change. To impose some meaning on an urban milieu characterized by flux, Kunming cabdrivers in part draw upon traditional Chinese urban symbolism, in which the urban and the rural are separated by a liminal zone of vague unease. "The cab drivers label as dangerous those places that are liminal ones—what might be thought of as the limits, the boundaries and portals of the city—between urban and rural space, where marginalized people enter the city, or where new forms of consumption occur." Notar concludes that "liminality takes on specific contours within cities—certain peoples and places are marked as more liminal than others." Familiar categories—some relating to space, some to cultural outsiders (e.g., Muslims, the Miao minority, and peasants)—often date back to the nineteenth century. Notar suggests that these categories "represent an effort to try to assert an older conceptual order on the radically new physical and social landscapes of the city, to harness old scapegoats for new conditions."

Rodney Collins (Chapter 11) discusses intersection between cultural and social structures, the colonial legacy, and capitalism in his analysis of how everyday Tunis residents experience the city. Differently situated actors in Tunis experience and move through the city differently, as can be seen in attitudes and frequentation of the city's enormous variety of coffeehouses. One of Collins's interlocutors, a male government employee, produced an impromptu mental map of the city, spatial, social, and gendered at the same time: "He pointed out that not only are there [mixed gender cafés] and [men's cafés], but that these are also found in distinct neighborhoods. He suggested that the former are found in the *centre ville* or in the wealthy *banlieues* whereas the latter can be found in zones considered *sha'abiya* [popular or working-class]." The interlocutor "elaborated upon his distinction by invoking the adjective *nathif* [lit. clean, proper] that he associated with the *banlieues* in general. Further, *nathif* not only marks the difference between the coffeehouses in the *banlieues* and the *sha'abiya* zones, but also distinguishes those institutions that are reputed to be frequented by female sex workers . . . or by homosexuals."

Tunisia, writes Collins, has a strong progressive history of juridical women's rights. Its abortion and divorce laws are arguably the most progressive in the Middle East, and the country even surpassed parts of Western Europe and North America in some areas. The comment of another interlocutor invites the following reflection by Collins. The interlocutor,

> who was in his mid-forties, did not mark the contradiction that the *qahwa sha'abiya* poses: being both for the people in name and yet not for women in practice. His views suggested that the *qahwa sha'abiya* is an institution that naturally suits a categorical understanding of 'the people' that precludes and, at least in the context of the practical normativity of Tunisia, predates the full participation of women . . . Despite . . . enormous strides in women's juridical rights [since the 1950s], everyday practice suggests a rigidly entrenched gender division with one symptom being the distribution of bodies in Tunisian coffeehouses as well as across the urban landscape.

This gendering of urban space, usually unnoticed by men, contrasts with the cafés on the Avenue Habib Bourguiba, which since 2003 has been subjected to the gentrification process described by Zukin (Chapter 1). Here, the cafés evidence *mixité*, gender mixing, as opposed to the mostly male-dominated cafés of the nearby *medina* (old city) and the popular neighborhoods. But this seeming embrace of diversity is not without irony. Freed from official pricing regulation (as stipulated by an elaborate state categorization of coffeehouses), the *mixte* cafés of the Avenue Bourguiba charge twice to ten times the *sha'by* price for a cup of coffee. "It is evident that movement through the social and physical space of Tunis is shaped quite extensively by the flows of capital," writes Collins. "Where capital is at its densest, women are most visible although other class-based exclusions may be attenuated in the process."

Chapter 12 is a reflection by Keisha-Khan Perry on the tyrannies, both subtle and not so subtle, of gentrification. A project by a Brazilian development firm in Gamboa de Baixo, a class- and racially mixed neighborhood of Salvador, brings with it an apparatus of segregating walls. Concrete walls, argues Perry, index racial boundaries, segregation, and the illegalization of life among the urban poor in Brazil. Black and poor urbanites become entangled in a state-neoliberal matrix of exclusion, racialization, and disposability, both of their bodies and their dwelling spaces.

Perry's insights from Salvador both echo and implicitly critique those of Mike Davis, who recently wrote more generally on major trends in the class character of global urbanism. By 2030, Davis argues, 5 billion of the world's 8 billion people will live in cities. Whereas according to the United Nations the world's rural population has stabilized and will never increase, cities are growing by 60 million per year, and 90 percent the world's population growth over the next generation will be accounted for by the urban areas of developing regions. "This urban population explosion will be almost completely delinked—or 'disincorporated'—from industrial growth and the supply of formal jobs" (Davis 2004: 11). Moreover, "the urban informal working class is not a labor reserve army in the nineteenth-century sense . . . on the contrary, this is a mass of humanity structurally and biologically redundant to global accumulation and the corporate matrix" (Davis 2004: 11).

Whereas Davis arguably sees "disincorporation" and redundancy in this condition of urban peripherlization, for Perry (as for another Brazilianist, Holston 2007), urban peripherlization actually often produces awareness and resistance. In Salvador, it is black women who are most vulnerable to the structural trends that Davis describes and their local manifestations. Black women in Latin America and throughout the black diaspora, writes Perry, and in particular those who live in the poorest neighborhoods, have been regarded as de facto noncitizens, facing the constant threat of expulsion to the peripheries of cities such as Salvador. It is black women's participation and grassroots organizing that has ultimately stood in the way of this peripheralization of local people by neoliberal development projects, creating in the process new ways of conceptualizing citizenship which fundamentally critique deep legacies of race and citizenship in Brazil.

CONCLUDING THOUGHTS

Perry's chapter, like the others, poses fascinating questions in relation to urbanism and even larger issues. For example, what kinds of new citizenship are Salvador activists fashioning in their encounter with the forces of neoliberal urbanism? As James Holston argues in his recent book, *Insurgent Citizenship*, because most Brazilians have historically been excluded from the legal use of land, modern Brazil has been pervaded by "illegality," in which the primary modes of everyday life and urban settlement have been constructed as illegal by state and municipal institutions. By the late twentieth century, Holston argues, illegal urban dwellers began organized struggles to assert claims over their land. In the process, they began opening space in Brazilian society for more egalitarian, less clientelistic forms of citizenship (Holston 2007; Mitchell n.d.). The use of the term "insurgent" to describe this process, as anthropologist Sean Mitchell points out, is intriguing, given both that the engagements between urbanites and the state in Latin America and the rest of the world increasingly resemble military conflicts. Because we now live in the age of the so-called Global War on Terror, with its military operations on urbanized terrain (to borrow the phrase of the U.S. military; see Graham 2007), more and more parts of the world are subject to militarized depictions of the city as space of insurgency. How will the militarization of an old dichotomy, that between urban centrality and the urban margins, affect the form of cities, urban class structures, and citizenship in the future?

In what ways, moreover, are new spaces for citizenship connected to the rise of Asia and Latin America, and the decline of the older industrial economies of the Global North/West? As the chapters by Kanna, Tang, Notar, Moser, Collins, and Perry all suggest, urbanity emerges from the encounters and frictions between multiple spatializations and ways of being, racial, gendered, and class-based, among others. The urban sphere, in turn, becomes both the arena of state or capitalist domination (as in the case of Haussmann's Paris, the cases of Bilbao, New York, and Shanghai analyzed by Zukin, or the new administrative capitals in Moser's chapter); but the urban can also become the arena of insurgent and democratic citizenship, as in Salvador and Holston's São Paulo, and, with the Arab Spring that broke out during the time of this writing, in contemporary Cairo and indeed Tunis. What are the intersections and articulations between the macroprocesses of Global North/West deindustrialization and projects of imperial hegemony, the rise of the new multipolar world of "Asia Rising" and "Latin America Rising," and the microprocesses of insurgent citizenship and constructions of the insurgent in the Global South? To help us and other scholars ponder these questions, Chen and Magdelinskas, in their Epilogue to the book, have distilled a number of empirical and theoretical lessons from the set of diverse secondary cities assembled here. For example, they point to how some secondary cities represented here have provided new insights into local economic restructuring under either globalizing or de-globalizing dynamics, and into the spatial construction and manifestation of national identity, social imagination, and political exclusion that challenge our expectations.

Finally, a theme not touched in this book is that of the ecological impact of urbanism. The roaring tigers of China and the Arab Gulf are adopting American-style consumption patterns—indeed, Abu Dhabi and Dubai are two of the world's largest per capita consumers of fossil fuels (Kanna 2011). At this juncture, the seemingly unstoppable and explosive urbanization of the world poses serious questions of ecological impact, given the energy needs and habits of urban populations. Whereas it is a hopeful sign that China seems to be getting more interested in green development and technologies, it is arresting to realize that none of the cases discussed in this book has made serious efforts to address the ecological problems arising from urbanization. Whether in rapidly developing China, the Arab Gulf, or Southeast Asia, or in the "de-globalizing" cities of the North American Rust Belt, no one seems to be willing to transform nineteenth-century industrial habits of consumption and development. These, like many others, are questions for further research on global urbanism that we hope will

Select Urban Sites Around the Globe

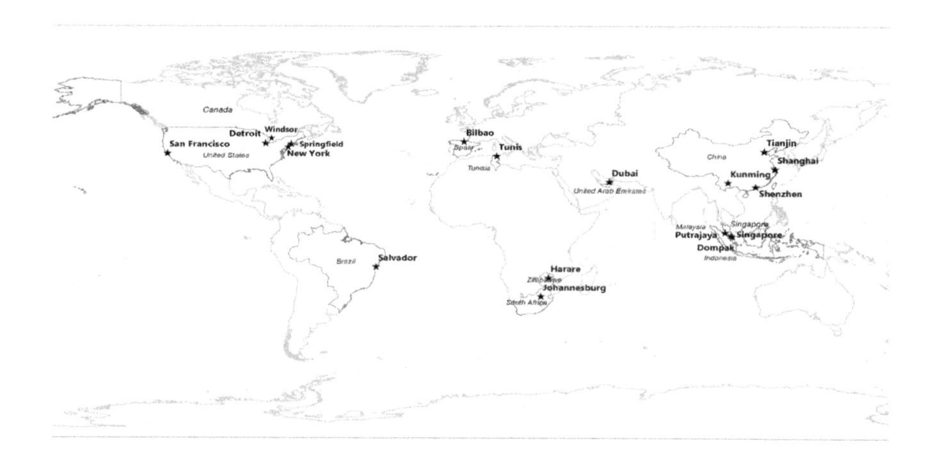

Map I.1 Locations of the cities covered in this book.
Source: Drawn by Nick Bacon.

emerge from the following chapters. To help get at or close to the questions not addressed in the book, we expect you to find a few suggested steps forward in the Epilogue.

NOTES

1. There is now a large literature in anthropology on the fractured, heterogeneous processes of globalization. But relatively little focuses primarily on urban issues. See Ghannam 2002, Kanna 2011, Srinivas 2001, Zhang 2001, and Zhang 2010 for recent anthropological studies of globalizing urban areas.

REFERENCES

Broudehoux, Anne-Marie (2007). "Spectacular Beijing: The Conspicuous Construction of an Olympic Metropolis." *Journal of Urban Affairs* 29(4): 383—399.
Chen, Xiangming (2009). *Shanghai Rising: State Power and Local Transformations in a Global Megacity*. Minneapolis: University of Minnesota Press.
Davis, Mike (2004). "The Urbanization of Empire: Megacities and the Laws of Chaos." *Social Text* 22 (4): 9–15.
Dawson, Ashley, and Brent Hayes Edwards (2004). "Global Cities of the South: Introduction." *Social Text* 22 (4): 1–7.
Elsheshtawy, Yassar, ed. (2008). *The Evolving Arab City: Tradition, Modernity & Urban Development*. New York and London: Routledge.
Friedman, John and Goetz Wolff (1982). "World City Formation: An Agenda for Research and Action." *International Journal of Urban and Regional Research* 6(2): 309—344.

Ghannam, Farha (2002). *Remaking the Modern: Space, Relocation, and the Politics of Identity in a Global Cairo*. Berkeley: University of California Press.
Graham, Stephen (2007). "War and the City." *New Left Review* 44: 121–132.
Hall, Peter (1966). *The World Cities*. New York: McGraw Hill.
Holston, James (1989). *The Modernist City: An Anthropological Critique of Brasilia*. Chicago: University of Chicago Press.Holston, James (2007). *Insurgent Citizenship: Disjunctions of Democracy and Modernity in Brazil*. Princeton, NJ: Princeton University Press.
Jacobs, Jane M. (1996). *Edge of Empire: Postcolonialism and the City*. New York: Routledge.
Kanna, Ahmed (2011). *Dubai, The City as Corporation*. Minneapolis: University of Minnesota Press.
Kusno, Abidin (2000). *Behind the Postcolonial: Architecture, Urban Space and Political Cultures in Indonesia*. New York: Routledge.
Marchal, Roland (2005). "Dubai: Global City and Transnational Hub." Pp. 93–110 in *Transnational Connections and the Arab Gulf*, edited by Madawi Al Rasheed. New York: Routledge.
Mitchell, Sean T. (n.d.). Review of James Holston, *Insurgent Citizenship*. E-Misferica 6.2. Electronic document, http://hemi.nyu.edu/hemi/en/e-misferica-62/mitchell, accessed on May 17, 2011.
Sassen, Saskia (1991). *The Global City: New York, London, Tokyo*. Princeton, NJ: Princeton University Press.
Sassen, Saskia (2009). "Cities in Today's Global Age." *SAIS Review* 29 (1): 3–34.
Sassen, Saskia Saskia (2012). *Cities in a World Economy*. Fourth Edition. Thousand Oaks, CA: Sage Publications.
Short, John Rennie (2006). *Urban Theory: A Critical Assessment*. New York: Palgrave Macmillan.
Srinivas, Smriti (2001). *Landscapes of Urban Memory: The Sacred in India's Silicon Valley*. Minneapolis: University of Minnesota Press.
Tsing, Anna Lowenhaupt (2005). *Friction: An Ethnography of Global Connection*. Princeton, NJ: Princeton University Press.
Zhang, Li (2001). *Strangers in the City: Reconfigurations of Space, Power, and Social Networks within China's Floating Population*. Stanford, CA: Stanford University Press.
Zhang, Li (2010). *In Search of Paradise: Middle Class Living in a Chinese Metropolis*. Ithaca, NY: Cornell University Press.
Zukin, Sharon (1989). *Loft Living: Culture and Capital in Urban Change*. New Brunswick NJ: Rutgers University Press.

Part I

Urban Present and Past

How Culture, History, and Politics Shape Global and Local Cityscapes

1 Competitive Globalization and Urban Change

The Allure of Cultural Strategies

Sharon Zukin

With the rise of branding as the predominant marketing strategy of our time, cities of all sizes and regions try to distinguish themselves by establishing a unique place-identity as their urban brand. Competition is the motivating factor, for since the 1970s cities have increasingly vied for the attention of companies that could bring investments and jobs and for affluent visitors and residents who funnel their spending power into the local economy. Although public officials define their efforts in terms of growth and jobs, they are pursuing a moving target, for the underlying dynamic is propelled by the mobility of both people and capital. Time annihilates space, as Karl Marx wrote more than a century ago, and with ever faster jet planes and electronic communications, both tourists and businesses easily move from place to place. It is impossible to ensure their loyalty. But cities try to be "entrepreneurial" about economic development (Eisinger 1988; Harvey 1989). They offer subsidies to companies to locate factories and offices within their borders, hire famous architects to design flagship projects that will spur media buzz, and promote their reputation as "the biggest," "the best," or "the capital of the world" for one kind of activity or another. Yet the uneven reward structure of a "winner-take-all" economy (Frank and Cook 1995) reduces the chances that first- and second-tier cities will achieve equal results.

Some researchers think there is an inbuilt tendency for a city's competitive strengths and weaknesses to reproduce a path-dependency where the accumulation of decisions and resources over time reinforces locational advantages (e.g., Kloosterman 2004). But men and women make the individual decisions that strengthen or weaken a city's economic base. Corporate executives in transnational corporations tend to regard cities of the same broad type as interchangeable; they ignore local traditions and give little weight to specific characteristics of the people who live there (Kotler, Haider, and Rein 1993: 12). So the CEOs of U.S.-based companies locate a call center anywhere that folks speak English and will work for lower wages, whether it is Sioux City, South Dakota, Manila, in the Philippines, or San Juan, Puerto Rico. Government action—or inaction—also makes a difference. The state-driven economies of Europe and Asia concentrate resources in cities and regions where companies already dominate their economic sectors—a geographic rescaling (Brenner 2004) that leads to even stronger and more competitive centers of manufacturing around Guangzhou in southern China, of high-tech industry in Singapore and South Korea, and of film, TV and music production in London. The urban policies of national governments can make or break a second-tier city's efforts to overcome relative decline and rise from the ranks (see Liu and Chen, this volume).

Just as with universities and football teams, widely publicized ratings dramatize status differences between cities. Only the biggest and richest cities of Asia, the United States, and Europe place high on the 2010 Global Cities Index, and the top four cities—New York, London, Tokyo, and Paris—get the highest rankings every year.[1] These ratings reflect strengths that second-tier cities cannot create on their own, for the top-ranked global cities are centers of business activity, especially financial trading; hubs of immigration and university education; nodes of corporate media headquarters; clusters of major cultural venues like museums, theaters and concert halls; and concentrations of embassies, consulates, and high-level conferences. This is the economic, social, and cultural infrastructure that makes and remakes a modern city's wealth.

Second-tier cities compete for other ratings. Sometimes these rating systems seem more humane or more people-centered than the global city rankings. According to *Forbes* magazine, for example, Pittsburgh is the "most livable" American city of 2010 (Levy 2010). The *Forbes* ratings emphasize low unemployment and crime rates, income growth, universities, and leisure facilities—qualities that may make a smaller city more pleasant than a global city but hardly distinguish it. *The Economist* uses still other criteria—social stability, health care, physical infrastructure—to rate "livable" cities around the world and comes up with a different list, headed by Vancouver.[2] Here too, though, as in the Global Cities Index, the top four cities—Vancouver, Vienna, Melbourne and Toronto—have won the top places for years.

Cities do not take a lesson from this experience. Ratings themselves exert pressure on public officials to keep up the competition with other places (McCann 2004). Local governments continue to pursue signs of distinction that they hope will help their cities stand out in the crowd. It's not just the "If you build it, they will come" approach popularized by the U.S. film *Field of Dreams* (1989). With globalization making competition more intense, all cities try to create "cultural experiences" that will attract "international residents and travelers," as the Global Cities Index says, leading to an unintended homogenization that hardly distinguishes them at all (Evans 2003). At least, this homogenization hardly distinguishes the "experience" of one first- or second-tier city from another. Despite local differences, homogenization reflects the strategic visions of urban growth that are shared by people with the cultural power to dream them up and the economic and political power to impose them.

Although competitive development strategies may not bring the desired results, they do change a city in two important ways. "First they provide a particular type of urban environment where the work of globalization gets done and second they provide a specific kind of global image" that is readily understandable as self-promotion (Marshall 2003). In other words, strategies aiming to transform a second-tier into a first-tier city really do create new urban spaces for work and play, and the spaces themselves—their names and prices, their architectural forms, and the lifestyle of modern luxury that they promise—"rebrand" the city in a different way. Play space is just as important as work space, for besides new corporate headquarters the iconic architecture of transnational capitalism today offers huge opportunities for consumption: indoor shopping malls, spectacular modern-art museums, and all-weather sports stadiums with expensive seats (Sklair 2009). Although these symbols of growth inspire pride, they push local governments beyond their fiscal limits and push local residents toward adopting modern, globalized modes of acquisition and display.

Often building new urban spaces also displaces longtime residents from their homes and from the streets where they seek a modest livelihood as artisans, traders, and vendors. Despite—or even because of—these evictions, the new districts look safer and less strange to outsiders' eyes (Graham 2004; Harvey 2008; Mitchell 2003). This is especially true in the centers of cities where dilapidated apartments, working-class cafés, and street markets are demolished, redeveloped, and "upscaled" to build new luxury housing, hotels, and corporate complexes, creating both a new global space and a new urban brand.

WHY CITIES LOOK ALIKE

Smaller cities do not want to be excluded from these global games. If they don't spawn major corporate headquarters or draw rich art lovers to auction houses and museums, they compete for a place on the global cultural circuit by developing more modest art fairs, film festivals, and even Cow Parades where painted fiberglass cows, or bison, or moose—depending on a city's chosen symbol—are installed on the streets as public art[3] (Bradley and Hall 2006). Other events promote local cultural industries. More than 150 cities from New York to Rio de Janeiro hold an annual or a semiannual fashion week, and design festivals for the furniture trade stretch from London to Ljubljana (Khemsurov 2008; Wilson 2008). Because modern-art museums have replaced factories as a symbol of collective wealth as well as a source of pride, every city wants one (Hamnett and Shoval 2003).

Keeping ahead of the competition is expensive, though, and officials of even the biggest cities complain that they can never do enough to maintain their city's lead. "We see ourselves as being in a competitive race with other cities from around the world," Daniel Doctoroff, the former New York City deputy mayor for economic development, said to a meeting of high-level cultural administrators and CEOs. "Many of [them] are trying to copy us, whether deliberately, or in some cases, unintentionally. . . . They're stealing our cultural institutions. There's a Guggenheim all over the world now" ("Creative New York" 2006).

But cultural competition is not the only way to explain the overwhelming force of standardization in global cities and would-be global contenders. In the early 1960s, the urbanist Jane Jacobs (1961) blamed the modernizers who worshipped progress and planned to rebuild all cities with right angles and straight lines. As architects and urban planners, they developed the intellectual tools and aesthetic styles that resulted in monolithic superblocks and high-rise towers, creating what Jacobs calls "the great blight of dullness." Alternatively, from an economic point of view, the geographer David Harvey (2001) sees the homogenization of cities resulting from the actions of investors, who tend to withdraw capital from one area or type of investment and shift it to another in a concerted effort to maximize profits. If developers can make more money and provoke less political control by building ranch homes in the suburbs, they'll do so; but when that becomes too difficult or costly, they'll switch to building loft apartments downtown. Today's concerted development strategies are shaped by the priorities of global investment funds that have grown enormously in recent years (Gotham 2006). Whether they are sovereign funds of foreign countries or transnational real-estate investment trusts, they target an increasingly wide range of buildings and construction projects, all tending to make the centers of cities more expensive and drive poorer residents to the periphery.

Like everyone else, investors, developers, and local government officials are influenced by the flow of ideas and strategies that travel around the world—"traveling ideas," the urban planning researchers Malcolm Tait and Ole Jensen (2007) call them. These ideas originate as a response to specific conditions in one city or another, but when they are talked about at international meetings or in the media, they stir people's interest everywhere (see Masiya, this volume, on two African cities adopting participatory budgeting from Porto Alegre, Brazil). When they are applied to one city after another, though, they lead to the homogenized landscape of "McGuggenization" (McNeill 2000). Often developers and public officials choose a competitive response that promises to outdo the efforts of other cities, like building the tallest skyscraper in the world. This works until a developer in another city builds a taller tower. Another strategy that has been used from Brooklyn to Beijing is to hire superstar architects to design eye-catching buildings and new urban districts (on Frank Gehry's plan for Atlantic Yards in Brooklyn, see Confessore 2006; on Beijing, see Ren 2011). Again, this works until developers or public officials in the next city hire the same superstar architects to reproduce their signature style—or until developers run into financial problems and scale down the architects' ambitious plans (on the scrapping of Frank Gehry's plan, see Bagli 2009). The risk is that these strategies will set up a treadmill of competition or an "iron cage" that guarantees a duplication of effects; this will in turn create an unwanted homogenization (DiMaggio and Powell 1983; Harvey 1989). Yet there is occasionally some nationally distinctive design of important buildings in the transnational architectural market of styles and practices (see Moser, this volume, on how this plays out in the new, second-tier Malaysian and Indonesian cities).

Like the aspirational consumption of shoppers who buy Louis Vuitton bags and Chanel sunglasses to express the high status they desire, business elites and political leaders engage in aspirational production, producing more modern-art museums, arts festivals, and cultural districts—all because they want to be *different.* They want to be different from other cities and help their city to escape its image problems (Greenberg 2008).

The value of aspirational cultural strategies is nearly always calculated in financial terms: increases in property values, in business and tax revenues, or even in charitable donations. When the artists Christo and Jeanne-Claude installed hundreds of bright orange flags in Central Park in the middle of winter in 2005, New York City tourism officials estimated that four million visitors came to see *The Gates* and bought so many souvenirs to benefit local nonprofit arts and environmental organizations that they poured $250 million into the city's economy.[4] For this reason, the officials and the media judged the event to be a great success—although Central Park was already attracting more visitors than any other city park in the country.

Despite the beauty of the installation, *The Gates* did not confirm New York's uniqueness. It was only one of Christo and Jeanne-Claude's many well-known projects that the artists have created since the 1970s by wrapping huge swathes of fabric around notable sites, from Berlin's Reichstag to Sydney's Little Bay. *The Gates* was a high-class variation on the Cow Parade, a way to join the Christo brand with the Big Apple of New York City. But the quantifiable success of *The Gates* encouraged the city government to repeat the experience by sponsoring more temporary installations of large-scale public art. In summer 2008, the East River sprouted waterfalls designed by the Scandinavian artist Olafur Eliasson.[5] Later that same year—in a transparent excuse for corporate branding—the Chanel Mobile Art Container, an exhibition featuring Chanel products in a temporary structure designed by the architect Zaha Hadid, was installed in Central Park on its way around the world, stopping in Hong Kong, Tokyo, London, Moscow, and Paris as well as New York City.[6] Again, these events did not establish or confirm New York's unique identity; neither did they promote the work of local artists. Instead, they put New York on a circuit of global and would-be global cities. *Not* to sponsor one of these installations would imply that the city had lost its place in the top rank . . . or so public officials thought.

These cultural strategies do bring several important benefits to local governments. First, they suggest that all cities can be winners. Second, unlike old smokestacks and docks, they're clean. Third, like shopping centers and business improvement districts, they make people feel safe, and they create a sense of belonging to a broad public of consumers. *The Gates*, the Guggenheim Museum in Bilbao, and the Cow Parade, as Jean Baudrillard (1977) wrote about the Centre Pompidou in Paris, are a part of the "hypermarket of culture" that keeps people enthralled, "in a state of integrated mass." As a result, public art installations, modern-art museums, and festivals have become a pervasive and an influential part of cities' toolkits to encourage tourism on the one hand and entrepreneurial creativity on the other, all in the spirit of developing a new, modern, globally oriented public culture. The effect, though, is to homogenize "difference" and repress local, multivocal cultural languages by a single dominant rhetoric of "global urbanism."

We can sample how these issues play out in different cities by traveling from the cultural entertainment district of Times Square in New York to the Guggenheim Museum in Bilbao and the "creative hub" of 50 Moganshan Lu in Shanghai. All of these spaces form nodes in global cultural circuits, spurring imitation around the world.

Times Square

New York's Times Square boasts a worldwide reputation as a vigorous entertainment center featuring everything from Broadway shows and movie theaters to bright neon lights, moving billboard advertisements, and feverishly milling crowds at all hours of the day and night. In the early 1900s, because of its cosmopolitan aura, people called Times Square "the crossroads of the world" (Taylor 1991). For most of the twentieth century, Times Square was also the legendary center of American popular culture. Men and women gathered there to read the latest headlines spooling off the huge, electric Motogram sign, and in Times Square Alfred Eisenstadt took a famous photograph, published in *Life* magazine, of a sailor sweeping a young woman off her feet and kissing her to celebrate the end of World War II. Times Square continued to draw visitors, including families with children, through the 1960s (Berman 2006). But as a heavily traveled, carnivalesque area of the city, it also drew men and women looking for casual encounters with strangers, often involving the rough trade of sex and drugs (Chauncey 1994). Because private real-estate developers showed no interest in building in Times Square, the city government proposed different kinds of projects, none of which attracted much attention—until the 1980s, when financial investors decided to build skyscraper offices that would erase the district's historic roughness and neon glitter as well as its seediness. This plan stirred so much opposition that the project was delayed for several years, until potential financing disappeared in the stock market's decline after 1987.

Much of the opposition came from the men and women who supported preservation of historic buildings and districts and led the city's elite cultural institutions. Spearheaded by the Municipal Art Society, opponents campaigned to protect the historic authenticity of Times Square by aesthetic means: by preserving its *look*. They proposed a law requiring each building in the district to wear a large illuminated sign on its façade—making each building, in effect, a giant electronic billboard. Broadway theaters, threatened with being sold by those who owned the land on which they stood and swallowed up by new office development, inspired another new law. This law enabled theater owners to sell "air rights," important for the opportunity they offer to build a much taller building on the site, to owners of nearby properties. The sale of air rights and their transfer to nearby sites would protect the concentration of low-rise theater buildings on the side streets while skyscrapers rose on the broad avenues around them, beginning on Broadway itself. At the same time, the New York City Police Department began vigorous sweeps of the sidewalks and porno shops, removing the pimps, hustlers, and hyperventilating, overwhelmingly male customers who made 42nd Street after dark so menacing, especially to women. For added force, the city government condemned properties occupied by porno shops that refused to change their stock or close, using the law of eminent domain to seize them "for the public good" and sell them to real-estate developers who would replace them with less troublesome businesses (Sagalyn 2001).

The combination of "aesthetic legislation," more vigorous policing, and use of eminent domain prepared Times Square for a change of public culture, one that relied on the Disney Company (Zukin 1995). The New York architect Robert A. M. Stern was the human connection between these strategies, for he served on Disney's corporate board and also supervised a local planning study of how to preserve the Broadway theater district. Stern recalls noting, when passing through Times Square one day in the late 1980s, that many people, especially women, came to matinee performances at the area's theaters by chartered bus from their homes in the suburbs. He then thought up—so he recalls—the idea of building attractions that would make Times Square safe enough for these suburban visitors to feel comfortable there. By chance or not, in the early 1990s the Disney Company was also thinking about expanding its business by producing shows for the commercial theater that would feature Disney film characters and theme-park rides. Placing Disney on Broadway, then, was an attractive prospect for both New York's urban planners and the Disney Company.

This suggested that there could be a healthy synergy between the family-oriented popular culture of Disney entertainment and the moral renewal of Times Square that the city government desired. With new laws protecting space in the district for commercial Broadway theaters, private developers began to plan new entertainment facilities for corporate tenants who wanted to be near the Disney beachhead. At that time, when Disney still seemed an invincible corporate giant with universal popular appeal, the company opened a Disney store on Times Square and renovated a historic theater, the New Amsterdam, next door, for the premiere of *Beauty and the Beast*. These attractions were soon joined by the studios of the cable music network MTV and a slew of themed restaurants, bars, and consumption spaces with well-known brand names—from Madame Tussaud's Wax Museum to B. B. King's Blues Club and a Hello, Kitty store.

By the usual financial measures, the new Times Square has been a rousing success. Every year, 38 million visitors come to gawk at the bright lights, eat at the restaurants, shop at Toys 'R' Us, and, if they can afford $100 tickets, go to the theater.[7] Almost 400,000 pedestrians walk through the area on a single day. The district's impact, according to the Times Square Alliance, a local business group, equals the combined economies of Bolivia and Panama. Entertainment is not the only type of business represented there. Although the cultural attractions bring crowds, Times Square is ringed by corporate office towers for the two envy-provoking, job-providing sectors of the city's symbolic economy: media and finance. But these types of business are most vulnerable to global financial downturns. The headquarters of Condé Nast, the Hearst Corporation, and the *New York Times* are at Times Square, and also the headquarters of NASDAQ, and the late, lamented offices of Lehman Brothers and Morgan Stanley, as well as those of many corporate law firms.

Cultural strategies of redevelopment are not risk-free. Media and entertainment venues may keep Times Square bright despite financial retrenchment, or they may decline as they did during the Great Depression, when the commercial theaters of 42nd Street degenerated into burlesque houses and cheap movie theaters. Moreover, the choice of cultural venues is important. Since the 1990s the seediness of the old Times Square has been replaced by a franchised, homogenized, plastic entertainment culture. Critics have complained bitterly of "Disneyfication"—a bland, mind-numbing sameness of corporate brand names, bad food, and banal entertainment. Branches of transnational businesses like Madame Tussaud's deny the unique location, detaching it from the city's history and from the area's pop culture roots in live performance. Despite pulsating electronic signs that make the asphalt ripple with expectation, Times Square has become a fast-food franchise zone of popular entertainment. It represents a standardized, global consumer culture in which the televised dropping of a glittering ball from the top of the Times Tower on New Year's Eve has become a cliché.

With the support of city government and a public-private business-improvement district, "Times Square" has become a brand whose name is used around the world, from Times Square Shopping Mall in Hong Kong and Saigon Times Square skyscraper in Ho Chi Minh City to Times Square Shopping Center in Dubai. Regardless of its real form in midtown Manhattan, "Times Square" suggests the glamour and profits to which any city may aspire.

Guggenheim Bilbao

Since the 1960s, when the French government created the Pompidou Center of Modern Art in Paris, modern-art museums have played a central role in three different cultural strategies. They are planned to be magnets of urban redevelopment, symbols of national prestige, and energizers of human capital for creativity and innovation. The new museums are nearly always located on geographically central land, often near the waterfront, that has lost its industrial uses and its economic value; the acquisition of this land is financed by the state. In Paris, the state located the Centre Pompidou in the Beaubourg area, a *terrain vague* of low-rent, dilapidated housing, small workshops, and cheap shops between the not-yet-gentrified Marais and the not-yet-demolished wholesale food markets of les Halles. After organizing an international competition to find an architect, the state-appointed commission chose a startling industrial design that placed air ducts and passageways on the museum's exterior, reversing the relation between internal structure and façade. Painted in bright colors, the exposed structural elements were meant to show the world that France would develop a set of progressive cultural programs and institutions that would propel Paris (and, by extension, the French nation) to a dominant position on global, modern art and music circuits. It was not only the design that challenged the European tradition of housing national art collections in monumental palaces. Awarding the commission to foreigners—the young British architect Richard Rogers and the Italian Renzo Piano—telegraphed the client's, in this case the state's, apparent new willingness to support innovation in all fields.

When the Pompidou's cultural strategy was copied by other clients around the world, it inflated the reputations of the architects who won the most prestigious museum competitions. They became "starchitects," their often-repeated signature styles became their "brands," and their practices grew and were globalized (McNeill 2008). By the 1990s, whenever business elites and political leaders in any city from Bilbao to Hong Kong announced plans to build a large cultural district on the waterfront, it was clear that these cultural facilities and their famous architectural designers were supposed to help the cities to achieve, or retain, a top-tier status.

Unlike the Pompidou Center, which is owned and run by the French government, Guggenheim Bilbao is an outpost of the global chain of the New York–based Guggenheim Museum. Like the Pompidou, however, Guggenheim Bilbao has local uses. It was conceived as an attention-getting symbol of business and political leaders' triple goal: breaking loose from declining industries, cleaning the abandoned remains of steel mills and shipyards from the waterfront, and advancing economic development as a bloodless counterterrorist campaign against Basque separatists. Like the Disney Company before it entered Manhattan, the Guggenheim was already primed for expansion. Since the mid-1980s, the museum's director Thomas Krens had planned to open branches in other cities of the world as well as sponsor a new Massachusetts Museum of Contemporary Art (MassMoCA) in North Adams, a declining industrial town in the Berkshire Mountains of New England. The Guggenheim Museum owned more artworks than it had exhibition space to show them in New York; many recently acquired new works and installations were so large they required galleries of their own (Zukin 1995).

Like Disney's role in the redevelopment of Times Square, the revitalization of Bilbao's industrial waterfront proposed a synergy between the economic strategies of Basque business leaders and urban planners, on the one hand, and the Guggenheim's cultural strategies, on the other. As in other new cultural districts around the world, the art museum would be joined by new shops and upscale apartment houses that would raise property values. But unlike in most other cities, Bilbao's leaders developed a broader, more ambitious strategy to modernize urban infrastructure by organizing international architectural competitions for the design of subways, bridges, and airport terminals. The main goal, though, was similar to the one that the French had defined with Centre Pompidou: to join the best of global design and local fabrication to craft a set of new urban symbols that would propel Bilbao—in competition with Madrid and Barcelona—to top-tier status. This represented an extraordinary break from the inwardness of cultural traditions in the Basque region toward an outward-looking global economy. Whether the museum would benefit unemployed local residents and artists, however, would remain an open question (Vicario and Martinez Monje 2005; Zulaika 1997).

In both its architectural design and the selection of art inside its galleries, Guggenheim Bilbao challenges traditional Basque identity. The architect Frank Gehry intended the curved, silver-color titanium panels of the museum's façade to represent the waves of the Nervión River flowing nearby and to suggest the toughness of the old waterfront's blast furnaces. But the museum's sinuous shape and flashy reflections of light imply a cosmopolitan glamour—the glamour of the actress Marilyn Monroe, as Herbert Muschamp, the late architecture critic of the *New York Times*, said when he visited the museum at its opening in 1997. The building suggests an "American style of freedom," Muschamp (1997) wrote. "That style is voluptuous, emotional, intuitive and exhibitionist." Nothing could be farther from traditional values in Bilbao, a city that in the past rejected both the minimalist steel sculptures of the American artist Richard Serra and the work of modern Basque artist Jorge Oteiza (Zulaika 2001).

Because the shipyards and steel mills were still standing when Muschamp visited Bilbao, it should have been easier for him then to see that the choice of titanium for the museum's façade referred to the city's old industrial power. But Frank Gehry translated the industries that had supported so many men and women into an abstract form, and the museum replaced a factory that, by some accounts, was still in working order. The new cultural district dramatized how finance and culture had come to occupy the dominant position in the city, with the Guggenheim both overlooking the city center and taking the major share of the regional government's cultural budget (Zulaika 2001: 12). Indeed, the museum has put Bilbao on a global circuit of both tourists and art. All of the exhibitions feature works by overseas artists, and most visitors to the museum come from overseas and other Spanish regions. This may be good for Bilbao because it opens the city to the world, but visitors tend to come once, stay only one or two nights, and leave without seeing the rest of the city (Lee 2007; Plaza 2000, 2006).

New cultural districts do attract media attention. Art magazines, travel guides, Web sites, and blogs promote new cultural scenes as places to go. Although the public may have no idea of what kinds of art, exactly, are showcased in the new facilities, they hear the buzz about them. Lower airfares, more free time, and a stronger emphasis on individual mobility encourage cultural tourism, especially among the affluent, highly educated men and women who likely collect such experiences as signs of their distinction. They are attuned to the emergence of new cultural districts and to all the new MOMAs and MOCAs around the world. They are also attracted to crowds. From the day it opened, the big open space in front of the Pompidou Center has been filled with tourists and buskers, young people and break dancers, many of them from overseas. In contrast, the big open plaza in front of Guggenheim Bilbao is generally empty except for tourists.

Business and political leaders in Bilbao have been praised for taking their future into their hands by welcoming change and revitalizing Basque culture. But exactly the same vision has shaped strategies in many other cities. From Melbourne to Manchester, hopes for cultural-led redevelopment have led to an endless series of waterfront cultural centers, cultural festivals, and cultural competitions. The European Union promotes these aspirations by funding the annual selection of a European Capital of Culture, which, like the Olympics, goads each city to rebuild its old industrial districts with flagship cultural projects. Asian cities develop their own flagship cultural projects to show they are no longer "cultural deserts" but top-tier contenders (Cartier 2008; Kong 2007).

Yet the repetition of these projects raises serious questions. Does it lead to the imposition of a single mold of creativity—one that excludes traditional artists? Does it exclude many local residents because they are neither producers nor consumers of art that wins the attention of global gatekeepers like art and lifestyle magazines or of overseas investors? And does it homogenize rather than distinguish cities? In the nineteenth century, industrial Bilbao was already integrated into the global economy. It was closely connected to European bankers and English factories that invested in its steel mills and shipyards. Today's modern-art museums are, in many ways, incubators of a new globalization. They are not just sources of pleasure and learning, but technologies of power that force urban cultures to open themselves to the outside, adapt to transnational markets, and become more cosmopolitan in a single, dominant "global" way.

50 Moganshan Lu

Like entertainment districts and modern-art museums, less formal artists' quarters—or "creative hubs"—show that a city is a global contender. This kind of hub is formed around the studios and live-work spaces of artists, designers, and other cultural producers whose presence rejuvenates an old industrial neighborhood or a complex of factory buildings. Before the 2000s, the cluster emerged spontaneously where artists gathered and opened galleries, performance spaces, cafés, and bars (Zukin 1982, 2010). Now, though, a creative hub is often subsidized by local government, trying to generate a spirit of innovation that could fuel a city's postindustrial economic development.

Many cities have the material requirements for these clusters to form: low-rent areas that are near to both financial and media firms where creative producers can sell their artistic or musical work, often on a freelance basis, and to low-price shops, bars, and restaurants, whose ethnic and working-class cultures provide a diversity of source material and a feeling of authenticity. From the viewpoint of young cultural workers, sharing the gritty streets with immigrants and workers helps to create a neo-Bohemia of creative energy (Lloyd 2006). In contrast to officials who want to attract a "creative class" whose members like hanging out in cafés and riding bicycles to work (Florida 2002), out-of-the-mainstream cultural producers really thrive on the jagged edges of uneven development, in areas that have not yet been sanitized or Disneyfied.

Although it is risky for local governments to support artists who may be critical of the dominant political party and the state, business entrepreneurs and political leaders in the rapidly growing economies of Asia have come to think that developing creative clusters may be useful as both an industrial and an urban policy (e.g., Kong 2009; Wang 2009). Beginning in the 1990s, a small number of these clusters formed spontaneously in Beijing and Shanghai more or less as they had done earlier in New York and London. Artists rented cheap studio space in factories that had been abandoned by manufacturers moving to more modern plants, sometimes in low-wage, rural regions of China.

In Shanghai, the artist Xue Song moved into studio space in a vacant, 1930s-era factory complex at 50 Moganshan Lu, near the Suzhou River, owned by Shangtex, a large textile and apparel holding company that had moved to a new development zone near the airport on the Pudong side of the river. More than one hundred artists were soon drawn to the vast amount of empty space on the Shangtex site; they were joined by cultural entrepreneurs from Europe and the United States who opened galleries of contemporary Chinese art, displaying work that until recently had not only been discouraged but was even banned by government and party leaders. The galleries now attracted foreign tourists and investors, who were as eager to "discover" new Chinese artists in gritty industrial surroundings in Shanghai as in similar surroundings in Williamsburg in Brooklyn or the East End of London. Shangtex became an eager patron—or landlord—of the artists' studios, connecting the idea of a creative hub with their own innovative technology for creating synthetic fibers.[8] Most important, the studios at 50 Moganshan Lu enjoyed the support of local party and government officials. On the one hand, Shanghai officials were influenced by the success of Factory 798, a complex of artists' studios and galleries that had opened in Beijing in 2001 and quickly spurred the development of a hip area of galleries, cafés, and boutiques. On the other hand, they saw that they had to compete with cultural strategies in Hong Kong and Singapore as well as in Beijing to keep Shanghai in the race for top-tier status. In 2002, the Municipal Economic Committee named the twenty-one-building complex an official industrial park; two years later, in line with national policies, this designation was changed to "art industrial park" (Wang 2006).

Building a creative hub at 50 Moganshan Lu suits Shanghai's ambition to become a global city that equals or surpasses its close competitors—Hong Kong and Beijing—as both a financial and a cultural center, a capital of the symbolic economy. Spaces are occupied by a variety of creative concerns: art galleries, graphic arts, architects', and design studios, and TV and film production facilities. Not surprisingly, M50's branding strategy capitalizes on language and images that originated in the United States and migrated to Europe and Asia: its slogan is "Suzhou creek/Soho/loft." As the hub's Web site states, these connections "embody that M50 is an integration of history, culture, art, vogue [sic], and originality." A meeting ground between the old loft buildings of Manhattan's SoHo and the new titanium marvel of Guggenheim Bilbao, Moganshan Lu is intended to upgrade the polluted waterfront, demonstrate the advantages of adaptive reuse over demolition of old buildings, and join technological innovation with artistic creativity. To justify the effort, M50's management evokes not just the appearance but the *experience* of authenticity: "The shabby factory buildings contain certain value, because the naked steel structure as well as the old brick walls and the mottled concrete make people feel the trueness and perfection of being existent."[9] These are high ambitions for a cultural strategy of urban development.

Moganshan Lu is popular with artists and tourists from overseas as well as from different regions of China. Visits are recommended by many guidebooks and Web sites, especially those aimed at the art world. Although there is just a small café and very few shops, several hundred visitors may browse through the galleries in a single day. Some of the artists who show their work there come from Hong Kong and Taiwan; they find the cost of living lower on the mainland, and in Shanghai they have access to an international market. Although a portion of the art is traditional in style, much more is almost shockingly modern and ironic, offering recognizable parodies of Maoist-era images and artifacts and suggesting that some of the more blatant examples of today's booming consumer society in China are as grotesque as the bourgeois capitalists caricatured by the German Expressionists in the early 1900s.

But despite its place on the global circuit of art tourists, 50 Moganshan Lu may not be successful enough to counter Shanghai's aggressive demolition of old buildings and districts. Although the city government plans to clean the pollution from the Suzhou River and develop a pedestrian-friendly green zone along its banks, construction crews continue to work day and night throughout the center of the city, tearing down factories like 50 Moganshan Lu and destroying the streets and houses that grew up around them during the first half of the twentieth century. These structures are being replaced by new, mixed-use districts of offices, luxury housing, and shopping, hardly places for "creative" work. Neither does the survival of Moganshan Lu guarantee permanent studio space for local artists. Rents there have gradually risen, and although they are still far less expensive than in New York or London, some artists have moved out of 50 Moganshan Lu and other creative hubs in search of cheaper space in the countryside (Wang 2009). In Beijing, high rents and real-estate development have had a similar effect on Factory 798 (Benaim 2006).

Today Shanghai offers a full range of cultural strategies with the city's own versions of Times Square and Guggenheim Bilbao as well as creative hubs. MOCA Shanghai, the city's first private, nonprofit museum, is located in a symbolically important place, People's Park, in the very center of the historic downtown. One of its exhibitions in 2011 featured fashions by Chanel, an interesting connection to the Chanel Mobile Art Container in Central Park three years earlier. Bridge 8, a pair of converted factory buildings in Luwan, offers architecture and design offices and graphic-arts studios, many connected to transnational firms. Nearby Tianzifang on Taikang Lu is a narrow street filled with storefront art galleries, upscale artisans' shops, and cafés, with artists' studios on higher floors. Yifei Originality Street, near the Science and Technology Museum in Pudong, is a new entertainment zone. Like the Disneyfied Times Square, it has "theme pubs, restaurants, art shops and nightclubs" along with "landscape lighting on trees and walls, with a central plaza" (Yang 2007). All in all, "originality" in the scenario of competitive globalization is hard to find.

LOOKING FORWARD

New York, Bilbao, and Shanghai have different types of market economies and governance systems. Their cosmopolitan cultures are rooted in different histories of migration and immigration; colonialism and postcolonialism; and democracy, fascism, or communism. But business and political leaders in all three cities, as in so many other cities around the world, believe that culture will help them to create a new urban brand for the global competition of the twenty-first century. Measuring the concrete effects of their cultural strategies is not an easy task. Counting the number of tourist visits or tourist dollars is quantifiable; calculating creativity is not. Moreover, contemporary cultural strategies run a strong risk of homogenization. The more cities try to use culture to demonstrate their "difference," the more homogenized they become. Globalization may turn out to be a cruel Darwinian evolution, selecting certain elements of cities (as well as languages and musical traditions) for preservation while blending others into hybrid, fusion, or dominant "global" forms.

For city dwellers eking out their existence day by day, such changes in the palette of urban life may expand some resources while reducing other opportunities for material survival. If they lose their house, their market stall, or access to the streets, the financial compensation they receive will not make up for the social networks that are broken and the change in the city's authentic character. Is authenticity only a fiction spun for the aesthetic tastes of highly educated, mobile consumers of places? Or is it a set of historical overlays, unexpected encounters, and uneven development that creates a dense urban patchwork of cultural identities? The cultural strategies of competitive globalization produce an instantly recognizable corporate zone with cultural amenities for the discerning "global eye" of the transnational corporate class while gradually creating a new edge city for migrants, both low-wage workers from rural regions of the world and native-born city dwellers who have been displaced (see Notar, this volume, on the mistrust and prejudice between the marginal locals and outsiders in the rapidly changing urban context in China). It is too soon to know whether either of these places—the global city center or the global urban periphery—will impose an inevitable, irreversible homogenization.

NOTES

1. http://www.foreignpolicy.com/articles/2010/08/11/the_global_cities_index_2010, accessed September 7, 2010.
2. http://www.economist.com/blogs/gulliver/2010/02/liveability_rankings, accessed March 15, 2010.
3. http://www.cowparade.com, accessed April 14, 2009.
4. http://www.christojeanneclaude.net/major_gates.shtml, accessed April 14, 2009.
5. http://www.nycwaterfalls.org, accessed April 14, 2009.
6. `http://www.chanel-mobileart.com, accessed April 14, 2009.
7. http://www.timessquarenyc.org, accessed April 14, 2009.
8. http://www.shangtex.biz/en/, accessed April 14, 2009.
9. www.m50.com.cn, accessed April 14, 2009.

REFERENCES

Bagli, Charles V. (2009). "Developer Drops Gehry's Design for Brooklyn Arena." *New York Times*, June 4.

Baudrillard, Jean (1977). *L'effet Beaubourg: Implosion et Dissuasion*. Paris: Eds. Galilée.
Benaim, Henri (2006). *Rendering Modernity: 798, an Avant-Garde Art District in Beijing*, senior thesis, Department of East Asian Studies, Yale College.
Berman, Marshall (2006). *On the Town: One Hundred Years of Spectacle in Times Square*. New York: Random House.
Bradley, Andrew, and Tim Hall (2006). "The Festival Phenomenon: Festivals, Events and the Promotion of Small Urban Areas." Pp. 77–90 in *Small Cities: Urban Experience beyond the Metropolis*, edited by David Bell and Mark Jayne. Milton Park, UK: Routledge.
Brenner, Neil (2004). *New State Spaces*. New York: Oxford University Press
Cartier, Carolyn (2008). "Culture and the City: Hong Kong, 1997–2007." *The China Review* 8 (1): 59–83.
Chauncey, George (1994). *Gay New York: Gender, Urban Culture, and the Making of the Gay Male World, 1890–1940*. New York: Basic Books.
Confessore, Nicholas (2006). "Developer Defends Atlantic Yards, Saying Towers Won't Corrupt the Feel of Brooklyn." *New York Times*, May 12.
"Creative New York," April (2006). http://www.nycfuture.org, accessed June 25, 2006.
DiMaggio, Paul J., and Walter Powell (1983). "The Iron Cage Revisited: Institutional Isomorphism and Collective Rationality in Organizational Fields." *American Sociological Review* 48: 147–160.
Eisinger, Peter K. (1988). *The Rise of the Entrepreneurial State: State and Local Economic Development Policy in the United States*. Madison, WI: University of Wisconsin Press.
Evans, Graeme (2003). "Hard-Branding the Cultural City—From Prado to Prada." *International Journal of Urban and Regional Research* 27 (2): 417–440.
Field of Dreams (1989). Directed by Phil Alden Robinson.
Florida, Richard (2002). *The Rise of the Creative Class*. New York: Basic Books.
Frank, Robert H., and Philip J. Cook (1995). *The Winner-Take-All Society: Why the Few at the Top Get So Much More Than the Rest of Us*. New York: Free Press.
Gotham, Kevin Fox (2006). "The Secondary Circuit of Capital Reconsidered: Globalization and the U.S. Real Estate Sector." *American Journal of Sociology* 112 (1): 231–275.
Graham, Stephen, ed. (2004). *Cities, War, and Terrorism: Towards an Urban Geopolitics*. Malden, MA, and Oxford: Wiley-Blackwell.
Greenberg, Miriam (2008). *Branding New York: How a City in Crisis Was Sold to the World*. New York: Routledge.
Hamnett, Chris, and Noam Shoval (2003). "Museums as Flagships of Urban Development." Pp. 219–236 in *Cities and Visitors*, edited by Lily M. Hoffman, Susan S. Fainstein, and Dennis R. Judd. Malden, MA, and Oxford: Blackwell.
Harvey, David (1989). "From Managerialism to Entrepreneurialism: The Transformation in Urban Governance in Late Capitalism," *Geografiska Annaler. Series B, Human Geography* 71 (1): 3–17.
Harvey, David (2001). *Spaces of Capital: Towards a Critical Geography*. New York: Routledge
Harvey, David (2008). "The Right to the City. *New Left Review* 53 (September–October), http://www.newleftreview.org, accessed November 20, 2008.
Jacobs, Jane (1961). *The Death and Life of Great American Cities*. New York: Random House.
Khemsurov, Monica (2008). "Design on Tour." *T: The New York Times Style Magazine*, September 21.

Kloosterman, Robert, with Eva Stegmeijer (2004). "Cultural Industries in the Netherlands—Path-Dependent Patterns and Institutional Contexts: The Case of Architecture in Rotterdam." Pettermanns *Geographische Mitteilungen* 148 (4): 66–73.

Kong, Lily (2007). "Cultural Icons and Urban Development in Asia: Economic Imperative, National Identity, and Global City Status." *Political Geography* 26: 383–404.

Kong, Lily (2009). "Making Sustainable Creative/Cultural Space in Shanghai and Singapore." *Geographical Review* 99: 1–23.

Kotler, Philip, Donald H. Haider, and Irving Rein (1993). *Marketing Places: Attracting Investment, Industry, and Tourism to Cities, States, and Nations.* New York: Free Press.

Lee, Denny (2007). "Bilbao Ten Years Later." *New York Times*, September 23.

Levy, Francesa (2010). "America's Most Livable Cities." http://www.forbes.com/2010/04/29/cities-livable-pittsburgh-lifestyle-real-estate-top-ten-jobs-crime-income.html, accessed January 28, 2012

Lloyd, Richard (2006). *Neo-Bohemia: Art and Commerce in the Post-Industrial City.* New York: Routledge.

Marshall, Richard (2003). *Emerging Urbanity: Global Urban Projects in the Asia Pacific Rim.* London: Spon.

McCann, Eugene J. (2004). " 'Best Places': Interurban Competition, Quality of Life and Popular Media Discourse." *Urban Studies* 41 (10): 1909–1929.

McNeill, Donald (2000). "McGuggenization: Globalisation and National Identity in the Basque Country." *Political Geography* 19: 473–494.

McNeill, Donald (2008). *The Global Architect.* New York: Routledge.

Mitchell, Don (2003). *The Right to the City: Social Justice and the Fight for Public Space.* New York: Guilford Press.

Muschamp, Herbert (1997). "The Miracle in Bilbao." *New York Times*, September 7.

Plaza, Beatriz (2000). "Evaluating the Influence of a Large Cultural Artifact in the Attraction of Tourism." *Urban Affairs Review* 36: 264–274.

Plaza, Beatriz (2006). "The Return on Investment of the Guggenheim Museum Bilbao." *International Journal of Urban and Regional Research* 30 (2): 452–467.

Ren, Xuefei (2011). *Building Globalization: Transnational Architecture Production in Urban China.* Chicago: University of Chicago Press.

Sagalyn, Lynne B. (2001). *Times Square Roulette.* Cambridge, MA: MIT Press.

Sklair, Leslie (2009). "Iconic Architecture and the Culture-Ideology of Consumerism." *Theory, Culture and Society* 27 (5): 135–149.

Tait, Malcolm, and Ole B. Jensen (2007). "Travelling Ideas, Power and Place: The Cases of Urban Villages and Business Improvement Districts." *International Planning Studies* 12 (2): 107–127.

Taylor, William R., ed. (1991). *Inventing Times Square: Commerce and Culture at the Crossroads of the* World. New York: Russell Sage.

Vicario, Lorenzo, and Manuel P. Martinez Monje (2005). "Another 'Guggenheim Effect'? Central City Projects and Gentrification in Bilbao." Pp. 151–167 in *Gentrification in a Gobal Context*, edited by Rowland Atkinson and Gary Bridge. London: Routledge.

Wang, Jie (2006). "Shanghai SoHo—50 Moganshan Road," http://www.chinadaily.com.cn/citylife, August 29, accessed September 7, 2006.

Wang, Jun (2009). "Art in Capital: Shaping Distinctiveness in a Culture-Led Urban Regeneration Project in Red Town, Shanghai." *Cities* 26: 318–330.

Wilson, Eric (2008). "The Sun Never Sets on the Runway." *New York Times*, September 8.

Yang, Li Fei (2007). "Chen's Creative Cluster Opens." *Shanghai Daily*, October 10.
Zukin, Sharon (1982). *Loft Living: Culture and Capital in Urban Change*. Baltimore, MD: Johns Hopkins University Press.
Zukin, Sharon (1995). *The Cultures of Cities*. Oxford and Cambridge, MA: Blackwell.
Zukin, Sharon (2010). *Naked City: The Death and Life of Authentic Urban Places*. New York: Oxford University Press.
Zulaika, Joseba (1997). *Cronica de Una Seduccion: el Museo Guggenheim Bilbao*. Madrid: Nerea.
Zulaika, Joseba (2001). "Tough Beauty: Bilbao as Ruin, Architecture and Allegory." Pp. 1–17 in *Iberian Cities*, edited by Joan Ramon Resina. New York and London: Routledge.

2 The Trajectories of Two "Asian Tigers"

The Imperial Roots of Capitalism in Dubai and Singapore

Ahmed Kanna

In 2007, I was taken aback when I saw an ad for Jumeirah Beach Resort, a Dubai luxury hotel complex, occupying a full page of *The New York Times* magazine. The ad depicted the helipad of the Burj Al Arab luxury hotel. A helicopter had just landed. In the foreground, a tall, beautiful blonde woman in an elegant white dress, dark sunglasses, and shiny jewelry walked powerfully toward the reader. In the background were various bellhops, wearing the red blazer and bowtie uniform of the Burj staff. The bellhops, whose seemingly South and Southeast Asian extraction surveyed a fair sweep of the Indian Ocean, were carrying the wealthy European woman's luggage and following her, presumably to one of the fleet of waiting Bentley cars in front of the hotel. This vignette reminded me of a more famous image from around a century ago, which appears in the chapter on the Gulf in the 2006 book *A Hundred Horizons: The Indian Ocean in the Age of Global Empire* by the historian Sugata Bose. In this picture, pith-helmet wearing Englishmen ride on the backs of Indian and Arab coolies who, along with their human cargo, are forced to carry the Englishmen's luggage from boats in the Gulf onto the shore. The setting then was the visit by British viceroy Lord Curzon to his Gulf dependencies, specifically the Trucial Emirates (the current UAE, Qatar, and Oman), in 1903.

In this chapter, I compare Dubai and Singapore to try to grapple with the theory of neoliberalism and to connect it with what I think are its roots, at least in these two cases, in the imperial encounter (see Moser, this volume, on colonial legacies in the context of Southeast Asian urbanism). My main work so far has been on Dubai, work that led me to explore the history of the wider Indian Ocean. In conversation (both literally and metaphorically) with scholars who are, unlike me, experts on the histories and cultures of the Indian Ocean arena (e.g., Sugata Bose and Engseng Ho), I began getting interested in Singapore. With the exception of political scientists Roland Marchal and Martin Hvidt, and, in occasional papers, the anthropologist Enseng Ho, scholars have not really compared these two cities. This situation should be remedied for various reasons. Here I discuss one: Dubai and Singapore have a shared history. Both were born in the crucible of the British Indian Ocean Empire. Both were nodes in larger regional buffers against competitors to

British hegemony in the Indian Ocean arena: the French, Russians, and Ottomans in the Western Indian Ocean, the Dutch in the Eastern Indian Ocean. On both, the British foisted invented dynasties, monarchies that governed their respective domains in the mode of centralized absolutism. These were monarchs who enjoyed British protection, in return for which they signed exclusive treaties ratifying British economic and political domination. In modern times both became city-state political economies with great relative autonomy in their respective Arab Gulf and Southeast Asian contexts. Both have adapted neoliberal policies in broadly similar ways. Dubai and Singapore's parallel histories mean that we are not comparing random cities; yet their geographic distance from each other and obvious cultural differences would hypothetically support recent theories in my own field, sociocultural anthropology, about the relationship between urbanism and capitalism (e.g., Hoffman et al. 2006; Ong 2007), about which I elaborate in the conclusion. My aim here is to offer some comparative thoughts based upon my own research on Dubai, and my review of scholarly literature on Singapore. In the conclusion, I ask: what can older theories about capital, culture, and urbanism offer newer theories? I have no firm conclusions at this point. I will offer a few thoughts on avenues for further research.

CONTROL OF MARKETS: NEW CONSTRUCTIONS OF INDIAN OCEAN SOVEREIGNTY

It is no coincidence that among the rationales for early nineteenth-century British intervention in the Malay Straits and soon thereafter in the Persian–Arab Gulf, ending "piracy" was central. As Sugata Bose has put it, the early nineteenth century witnessed a new construction of Indian Ocean maritime trade: Western imperial states now saw themselves as beset by "pirates" (Bose 2006: 45). This gave an otherwise patently imperial project the patina of self-sacrifice and duty. As Lord Curzon, the viceroy of India, asserted in a 1903 address to the assembled sheikhs of the so-called Trucial Oman emirates (as the United Arab Emirates were known prior to independence in 1971):

> We found strife and we have created order . . . The great Empire of India, which it is our duty to defend, lies almost at your gates . . . We are not now going to throw away this century of costly and triumphant enterprise; we shall not wipe out the most unselfish page in history. The peace of these waters must still be maintained; your independence will continue to be upheld; the influence of the British government must remain supreme (Bose 2006: 37).

Until around 1820, when the British entered the region, the southern Arab Gulf was controlled by the Qawasim (sing. Qasimi), a maritime people from the emirate of Sharjah, located directly east of Dubai. To the British, the tolls charged by the Qawasim interfered with access to India, and they sent a "devastating naval expedition" against what was called, in colonial parlance, the "Pirate Coast" (Onley 2005: 30). They imposed the General Treaty of 1820, subjecting the rulers of the coast of Oman (which then encompassed today's UAE) to British India. These were followed by the Maritime Truces (1835–1853), which incorporated Abu Dhabi, Dubai, Ajman, the Qasimi state (today's Sharjah), and Umm al-Qawain—all states of the contemporary UAE—into the Empire as the "Trucial States" (Onley 2005: 31). At this time, the purpose of the treaties was to ban maritime warfare during the pearling season. The Exclusive Agreements (1880–1916) followed, granting Britain exclusive political relations with, and control of the foreign affairs of, the Trucial States, Bahrain, Kuwait, and Qatar (Onley 2005: 32). Along with securing access to India, these treaties created a buffer between the British zone and those of other great powers, such as France, Russia, and the Ottoman Empire, which, respectively, made inroads into Egypt and Iran (1798–1809), Central Asia and Iran (1868–1885), and the eastern Arabian peninsula (1871–1872). By persuading Gulf potentates to sign exclusive treaties with them, the British created "a screen" of principalities along the northwest boundary between India and the Middle Eastern–European region (Onley 2005: 42).

Although a less urgent concern for the British in Southeast Asia, "piracy" became a prominent justification for intervention by the 1830s. Until the mid-1830s, the Illanun and Balanini expeditions from Mindanao and the Sulu Sea, respectively, would disrupt trading activities in the Malay Straits. "Pirates" of the Riau-Lingga archipelago were also involved: "So intense was their rivalry that they often neglected their prey in order to fight with one another" (Ken 1991: 48). In 1836, the Indian government initiated a plan for suppressing Malay piracy, making gunboats and money available for counterpiracy measures. By the 1850s, the British had largely succeeded in eliminating "piracy" in the Malay Straits (Ken 1991: 48). This helped free Britain to elevate Singapore as an imperial node and a center of free-trade ideology and British expansion in the Malay region. Like the small principalities of the Gulf, Singapore was also to be a shield against competing imperial claims on a shared sphere of interest (the Malay economy), in this case, those of the Dutch and the French (Soon 1991: 371).

The 1819 Anglo-Malay Treaty formalized relations between Britain and the Malay authorities. It is interesting to note the echoes of the exclusive treaties imposed on the Gulf during in the late nineteenth century. In return for British "protection" against external enemies, the British were permitted to establish a "factory" (the contemporary term for a trading post). This was, moreover, to be an exclusive treaty, forbidding the Malay authorities to make deals with other foreign states (it is notable that these treaties were in this way similar if not identical to the rent deals the British made with Gulf rulers). The Temenggong, the local chieftain of Singapore, for example, received half of all levies on native vessels. In 1823, in another echo of the Gulf rent agreements, the British extended control over the entire island, with the Temenggong and the Malay sultan agreeing to receive allowances for life in exchange for renouncing rights to port duties (Chew 1991: 36–39).

The discovery of tin in the Malay Straits in the 1840s, like the example of piracy, betrays one of the real purposes of empire: control of markets. In the middle of the nineteenth century, argues Engseng Ho, tin mining was a collaborative enterprise between Malay and Chinese. Malay chiefs contracted out monopoly mining rights to foreign bidders. These bidders were Chinese, especially the Hokkien diaspora referred to by the literature as the "Amoy Network" (from Amoy, the name given by the English to Xiamen, a major Southeast Chinese entrepôt of the nineteenth century). Hokkien traders would organize themselves into conglomerates based upon shared language and economic interest. By the 1860s and 1870s, these so-called secret societies began competing with each other for control of mining rights, especially in Perak (in modern-day Malaysia). The British colonial archives represented these conflicts, collectively known as the Larut War, as simply based upon primeval hatreds. In fact, as Ho shows, these were conflicts over markets between Chinese conglomerates. With increasing demand for the tin can during the U.S. civil war and the opening of the Suez Canal in 1869 (shifting tin markets from China to the West), tin had suddenly become an enormously valuable commodity in the global economy. Britain wished to supplant China as the world's primary supplier of tin. Representing the Chinese secret societies as a source of chaos and the Malay chiefs as incapable of suppressing them became a useful pretext for the British to capture the Malay-Chinese political economy in the second half of the nineteenth century (Ho 2002).

Under The Pangkor Treaty, the sultan of Perak was bound to accept British advice "on all matters except those concerning Malay religion and customs" (Bose 2006: 51). The Pangkor agreement was soon extended to Selangor and Sungei Ujong, and, by the late 1880s, to Pahang and Negri Sembilan—that is, to most of the states incorporated into the British-invented Malay Federation. The treaty legitimized British rule of Malaya and criminalized competition to British dominance of Malay markets. A concept of "the government as absolute sovereign over civic space" emerged, a pattern applied elsewhere in British Malaya, including Singapore (Ho 2002).

Although the exclusive treaties between the British and, respectively, the Malay authorities and the Gulf sheikhs, are essentially identical, there are also, at least during the nineteenth century and the early part of the twentieth century, important differences between British Malaya and Trucial Oman. As Christopher Davidson has shown, the British treaty system isolated the Gulf and made it a desperately impoverished region until the successive oil booms between the 1930s and the 1970s (Davidson 2008). Emirates such as Abu Dhabi, Dubai, and Qatar were among the poorest, least developed parts of the Middle East until the 1970s. In the nineteenth century, Singapore became one of the premier entrepôts in the world, a bastion free-trade. Singapore's innovation was that it was a free port, something that Dubai would in fact emulate during the twentieth century. Indeed, among the strategies of the rulers of Dubai were their attempts to poach Gulf trade from rivals by adopting free-port policies throughout the twentieth century. Meanwhile, it should be added that, whereas imperial functionaries such as Curzon saw in the Qasimi and Sulu sultanates "the quintessence of an Islamic world whose activities centered about piracy and slavery" (Bose 2006: 45), another and less self-serving view of "piracy," as of "primeval" intra-Chinese conflicts at a later date, is as a set of competing claims to markets, which inevitably conflicted with monopolistic European trading practices in the imperial era (Bose 2006: 45). The British solution to both piracy and secret-society conflicts—the creation of pliable, centralized, autocratic states deploying a unitary, European-style sovereignty—would have a momentous impact on the Indian Ocean states (Bose 2006: 25).

THE EMERGENCE AND TRAJECTORY OF NATIONALISM

This impact is perhaps best seen in the intersection of nationalism and the postindependence paths of Dubai and Singapore. In the era of Indian Ocean Empire, the British sought to put in place pliable local rulers and to envelope them in the trappings of "tradition" and legitimacy. With respect to the Gulf emirates and the Straits Settlement, the British chose two variants of a policy embedded in the same ideology and vision of sovereignty: unitary, indivisible, "a major break from ideas of good governance and legitimacy that had been widespread in the Ottoman, Safavid, and Mughal domains and their regional successor states" (Bose 2006: 25). Moreover, and especially with respect to India after the Great Revolt of 1857, and later the Bay of Bengal and the Arab Gulf emirates, the British "juxtaposed with their own monolithic sovereignty a particularly fake version of sovereignty invested in reinvented 'traditional' rulers" (Bose 2006: 25).

Predictably, British domination benefited a few local actors and marginalized many others in both the Gulf and the Southeast Asian contexts. Benefitting in particular were Dubai's invented dynasts and Indian merchants (the so-called Banians), on the one side of the Indian Ocean, and Singapore's British-connected capitalist elite, comprised of Hokkien financiers (e.g., the so-called Babas of Penang and Singapore) on the other (Al-Sayegh 1998; Ho 2002). However, whereas the independence and postindependence periods witnessed broad continuity in Dubai, they saw a break in Singapore. This was primarily the result of the ways in which the two cities experienced the transition from imperial domination to independence in the period of the emergence of nationalism.

In the Trucial Emirates (the present-day UAE), according to Christopher Davidson, Britain's successful attempt to create a system of British-dependent dynasties resulted in the arrest of the "centuries-old ebb and flow of tribal power" (Davidson 2008: 18). I would go even further to maintain that British intervention arrested the very possibility of substantive politics in the UAE. During the 1950s, a nationalist movement inspired by Egypt's Gamal Abd al-Nasser and known as the Dubai National Front (DNF) emerged in the Trucial Emirates. British imperial archives from the 1950s evince the distaste if not paranoia on the part of the British resident and political agent vis-à-vis the "Nasserite threat." The new dynasts, such as Abu Dhabi's Zayed bin Sultan Al Nahyan and Dubai's Rashid bin Said Al Maktoum, shared this distaste. For example, Rashid routinely executed British desiderata, such as the deportation or arrest of nationalists and reformers. Both Zayed and Rashid, moreover, offered to personally pay for a continued British presence in their domains after London's decision to dismantle the empire east of Suez in 1968 (Davidson 2008: 39–54, 63).

At independence in 1971, the UAE state took the form Britain had intended: dynastic and autocratic rather than republican or constitutional, let alone even nominally socialist or anticolonial. Davidson puts it well: "The Political Resident was well aware that the Trucial states (along with Qatar) were relatively 'primitive and *needed to be kept under a more colonial character*,' and therefore recommended a much looser system that *still retained monarchical sheikhdom-level powers and respected local institutions*" (Davidson 2008: 59, emphasis added). The term "local institutions" is to be understood here more specifically as referring to political institutions of a hierarchical, unitary, and "tribal" nature, conflating "tribal consultation" with what is sometimes called the Gulf's "Arabian" version of democracy.

From a reformist republican perspective, the timing of UAE independence was unfortunate. The British role in ensuring that the southern Gulf states took a conservative cast should be contextualized in the post-1947 collapse of the British Indian and Near Eastern empires, evidenced by well-known events such as Indian independence, the rise of Nasser, and the 1956 Suez crisis, but also by less celebrated events such as the decolonization of Zanzibar and the overthrow of its invented dynast in 1963–1964 and British withdrawal from and the inroads of Arab nationalism into Yemen in the mid- to late-1960s (Davidson 2008: 59).

The path that the Singapore state took, by contrast, was the result of the coincidence of timing and war. If Britain had a relatively free hand in shaping the outcome of independence in the Arab Gulf, it had no such luxury in Southeast Asia. The Syonan, or Japanese occupation, years (1942–1945) were a bloody interregnum between a British-dominated prewar order and a relatively robust, independent republican postwar order in Singapore. An important legacy of the Japanese occupation was the puncturing of the image of an invincible British Empire. Lee Kuan Yew and his colleagues, along with their opponents, belonged to the generation that went through the occupation. Thus, it is interesting to compare, on the one side, the conclusion that Lee drew from the collapse of British and Japanese occupation, with, on the other side, the Emirati dynasts' desire for continued British protection. Lee, asserting Singapore's independent and self-sufficient postwar streak, reportedly said: "There was never a chance of the old type of British colonial system ever being repeated" (Thio 1991: 110–111). This is in striking contrast to the tendency among Arab Gulf states to adjust oil production rates to suit American market demands, engage in desperate scrambles to attract British and American military protection, peg their currencies to the U.S. dollar, and pander to Western expatriates, consumers, and tourists.

There is, however, a haunting similarity in Dubai's and Singapore's authoritarianism, what can, following the Singaporean opposition figure Soon Juan Chee, be summarized as a probusiness and antidemocratic ideology and state project (Chee 2001). The British, via their protégés, the Al Maktoum dynasts, succeeded through either expulsion or co-optation of the most troublesome nationalist elements in destroying Dubai's reformist tendencies. The Singapore People's Action Party of Lee Kwan Yew (PAP), by contrast, needed little help from the imperial overlords to destroy the opposition. Like their counterparts in the Gulf, the Singapore opposition tendencies wanted greater citizen participation in politics as well as accountability of their governments to the citizenry. But unlike in the Gulf, where dynastic state builders were avowedly reactionary in their politics, the PAP shared with its opponents certain reformist commitments such as distributive developmentalism, if in a politically defanged version under highly exclusive one-party rule. Nevertheless, whereas the Communists, feminists (e.g., the Malay Women's Welfare Association), and trade unions that were in the vanguard of Singapore reformism (Thio 1991: 110–111) were largely absent in the British-invented "tribal, monarchical, and primitive" UAE, in both Dubai and Singapore one notices the emergence in the independence period of rule by one faction (the Maktoum, the PAP) concentrating in itself all sovereign power.

There is indeed one more striking parallel. In both city-states, the discourse of the ruling faction borrowed heavily from that of its imperial overlords, representing all opposition tendencies as potential threats to national sovereignty or cultural authenticity. Moreover, such discourses seem to be peculiarly immune to changing historical circumstances. In Dubai, for example, present-day migrant worker agitation for better working conditions is seen and represented by the state as a threat, in almost exactly the same ways and with nearly identical terminology, as were perceived threats from nationalists and progressives during the 1950s (Kanna 2011). Early in the Singapore independence period, the PAP deputized its Internal Security Department (ISD) to seek and destroy so-called communists, a term that was applied to a wide variety of opponents, from the now-defunct but once influential Barisan Socialis (Socialist Front) party, to traditional Chinese schools, to members of the Catholic Church, to the opposition Worker's Party (Chee 2001; Englehart 2000).

THE CAPITALIST DEVELOPMENTAL STATE IN THE NEOLIBERAL ERA

Entrepôt Growth

EMAAR is one of Dubai's more impressive corporations and brands. Dubai's largest real-estate developer, it is the company that has produced many of the more famous Dubai projects over the past decade.[1] Its most prominent recent project is the building formerly known as the Burj Dubai (Dubai Tower), rechristened Burj Khalifa in 2009 (after Abu Dhabi's ruler, Khalifa bin Zayd Al Nahyan, bailed out Dubai following the financial collapse of 2008). The Burj is a centerpiece for the new downtown Dubai, an enormous urban enclave which was conceptualized during the past decade and continues to take shape. The mandate to shift Dubai's geography of centrality in such a rapid, dirigiste fashion has rested largely on the American-educated and Singapore-trained chairman of EMAAR, Muhammad Al Abbar.

As both Roland Marchal and Martin Hvidt have shown, since the 1980s Dubai has in fact borrowed lots from the toolkit that Singapore had established with such striking results after independence (Hvidt 2009; Marchal 2005; compare to two other contributions to this volume: Zukin's discussion of how cities borrow development strategies from each other, and Moser's analysis of "serial seduction" and of global Islamic urbanisms). In the 1980s, East Asian states such as South Korea, Taiwan, and Singapore came to exemplify what political scientists call the capitalist developmental state model (Hvidt 2009: 398–399). Following neither the Washington Consensus of unregulated markets nor the Santiago Consensus, a more eclectic statist model, the capitalist developmental state model is based upon "significant government interventions in the economy . . . a historical legacy of strong and economically active states, traditions of social and political hierarchy and strong nationalist sentiments underpinned by cultural homogeneity and reinforced by external threats" (Hvidt 2009: 399). Although perhaps more applicable to relatively large, culturally homogeneous nation-states such as Japan (seen by Hvidt to be the birthplace, in the 1920s, of capitalist developmentalism) and South Korea than to multicultural city-states such as Dubai and Singapore, the relevance of this model to both city-states is well supported by Hvidt. For example, according to Nabil Ali Alyousuf of Dubai's Executive Office (headquarters of the Executive Council, the body appointed by the ruler to conduct economic and urban planning in the city), Singapore became a guide for Dubai "especially in vision and proactive leadership" (Hvidt 2009: 399).

More specifically, Dubai, like Singapore (and Hong Kong), has pursued a path of "entrepôt growth," situating itself as a strategic node between Asia and Europe (Oh 2009).[2] In 2009, it was estimated that Dubai had become the sixth largest container port handler, rivaling Singapore's consistent position at or near the top (Oh 2009). Hvidt distinguishes this entrepôt growth from the path pursued by "most advanced countries," which have shifted from agriculture to nonagricultural sectors such as industry and services. "Entrepôts . . . lack the agricultural point of origin and develop large service and commercial sectors in line with their function as intermediaries between primary-exporting hinterlands and regional, imperial, and world economies" (Hvidt 2009: 400).

Geographically and economically, Dubai may be regarded as an entrepôt astride one of the more volatile boundaries of the post-1970s global economy. The independence of the UAE (1971) and the establishment of Dubai's first free-trade zone at Jebel Ali (1979–1980) bookend a number of events of regional and global significance. Oil was discovered in Abu Dhabi in 1958 and in Dubai in 1969. Soon after independence, the national oil industry was reconstituted, with Abu Dhabi acquiring a 60 percent stake in the country's largest oil company. This was followed by the 1973 OPEC embargo and economic booms in the Gulf, Iran, and Iraq. Between 1968 and 1973 Dubai's economy expanded sevenfold (Zahlan 1978: 185). Meanwhile, wars and coups convulsed the region. Lebanon was destroyed in a protracted civil war and by foreign invasions, the effects of which still afflict the country (enabling Dubai to take over Beirut's functions as a banking, finance, and tourism capital of the Arab world). Radical Islam filled the vacuum left by Arabism, nearly toppling Saudi Arabia's ruling dynasty and doing so to Iran's in 1979. Reacting to U.S.-fueled provocations, the Soviet Union invaded Afghanistan, also in 1979.

Dubai's significance to the region—not only the Middle East and the Indian Ocean, but also a transcontinental arena connecting Africa, East Asia, and Russia—began to increase at this moment. By the 1990s, the collapse of the Soviet Union and neoliberal restructuring in India reinforced connections with these regions. During the 1990s, for example, a large number of *chelnoki,* Russian itinerant traders, arrived in Dubai. Tourists and real-estate investors soon followed. In a recent estimate, there were about ten thousand Russian citizens living permanently in Dubai (Starostin 2007). From India and Pakistan came not only working-class migrants but also middle-class professionals, making Dubai home to one of the most prosperous Indian communities in the world (*The Economist* 2002) and also a platform for South Asian money laundering, with Dubai and Singapore, as well as Hong Kong, operating as offshore linkages in the *hawala* network (Janardhan 2007).

Iraq and Iran are also central to Dubai's modern evolution. Dubai became a destination for Iraqis, and a place where they increasingly invested their capital, after the Gulf War of 1991. These movements increased following the more recent American war on Iraq, and reexport from Asia, especially of Japanese stolen cars, goes through Dubai to Iraq (Bowcott 2002; Parker and Moore 2007).[3] In fact, even before the 2003 invasion, reexports from Dubai to Iraq were estimated to be around $550 million in 2002, from $176 million in 2000 (Marchal 2005: 96–99). Dubai, some scholars have also suggested, may be the "economic capital of Iran" (Marchal 2005: 96; see also Adelkhah 2001). In 2004–2005, 50 percent of reexports out of Dubai were going to Iran. Meanwhile, about seventy thousand Emirati nationals are of Iranian origin, and they, along with the approximately one hundred thousand Iranian migrants in the UAE, facilitate the operations of the three thousand Iranian companies in the country (half of which are based in Dubai). Since the revolution, the United States and Iran officially broke off relations, but continued to trade with each other by way of Dubai (Marchal 2005: 99). Other examples since the 1990s include Chechnya and Afghanistan. In the case of the latter, the Taliban was recognized by only a handful of states, the UAE among them. In 2000, Afghanistan was the tenth largest market for Dubai reexports (Marchal 2005: 99).

Although connections between Africa and the Arab Gulf go back to the period of the British Empire, the current reconstitution of Dubai as a central node along African networks is more recent, dating to the end of the Cold War (see Fenelon 1976 and Marchal 2005 for two different perspectives).[4] Restructuring in various parts of Africa, often as a direct result of the Soviet Union's collapse, guided the capital "freed" from these countries, increasingly, to Dubai. In the aftermath of September 11, 2001, Nigeria's elites, for example, shifted laundering activities from Europe and the United States to the Middle East by establishing front companies in Dubai (*International Oil Daily* 2007). Economic restructuring led to the emergence of new entrepreneurs who could not operate within the formal banking system and who looked for products cheaper than those in traditional European markets, products that are in plentiful supply in the duty-free malls of Dubai (Marchal 2005: 104–106). With the end of the Cold War, aid to African countries shriveled. By the 1980s and 1990s, the development differential between Asia and Africa, which was about equal in the 1950s, greatly favored Asia. All of this coincided with the expansion of Dubai, the celebrated boom of the Gulf studies literature, as the city became the interface between the two continents.[5]

"Baroque Ecology" and "the City-Corporation"

Singapore is often analogized as a corporation and so indeed is Dubai. But Dubai's demography (at least 80 percent of the population is nonnational), combined with its access to Abu Dhabi's petrodollars, means that it is an even more dramatic case of "city as corporation" than is its Southeast Asian counterpart. The lack of accountability of the ruler and the sixteen-member Executive Council (similar to the lack of accountability of the PAP) "is a testament to Dubai's particularly strong version of (Gulf states' common policy of popular cooptation) and its more business-focused national population that to some extent views the government as a board of directors rather than as a forum for political participation" (Davidson 2008: 159). What this means is that, by virtue of the accident of demography, Dubai's state, much like the unregulated corporation in the neoliberal West, ends up having no social obligations (education, healthcare, etc.) to most of its residents. This creates a "uniquely favorable situation," according to Martin Hvidt. "While other countries struggle to educate and especially reeducate their population as the countries pass through various stages of development, Dubai basically 'purchases' its workforce on the international market to suit current needs" (Hvidt 2009: 403). This allows for both flexibility in hiring and firing labor and quick strategic changes in relation to the changing demands of the global market (Hvidt 2009: 403). Thus, in Dubai, it is common for foreigners to get "runwayed" (deported) for various reasons, among them economic redundancy. Moreover, the threat of being runwayed, as several of my expatriate and migrant worker interlocutors attested during my field research in Dubai, ensures these foreigners' good behavior. Reports of migrant workers getting fired en masse during the recent financial crisis, while shocking, are predictable in a society and political economy narrowly structured to protect both the distributive welfare state claims of the small percentage of the population making up the ruling ethnic group (Longva 2005) and the rights of multinational capitalist actors (Abdulla 1984; Kanna 2011).

Since the 1990s, both Singapore and Dubai have adapted a set of strategies that can loosely be termed neoliberal. Here too there are remarkable similarities. The neoliberal turn is also a good study both in the role of ideology and symbolic values in shaping the policies and strategies of global cities, and in the ways in which such strategies, in turn, reshape local values.

The 1997–1998 Asian economic crisis is a watershed in Singapore's specific neoliberal turn, pushing the city-state to try to reposition itself as a hub for the global knowledge-driven economy (Ong 2007: 178). The government's agenda for the early twenty-first century was to diversify Singapore's strategy to complement manufacturing, especially in areas such as electronics, chemicals, and pharmaceuticals (Ho 2009: 79), with informational and technological production appealing to an increasingly interconnected global economy inflected by new alignments of localism and regionalism. This strategy envisions Singapore as an incubator for reterritorializing Asian business and a testing site for drugs and treatments targeted at Asian patients (Ong 2007: 182–183).

According to Ong, this has entailed two kinds of reconceptualization, pertaining, respectively, to the spatial imaginary of the state and to the ethical and symbolic dimensions of Singaporean identity, to what it means to be culturally Singaporean. With regard to the first, the city-state framing of the state's sovereignty has been reformulated along the lines of what

Aihwa Ong calls "Baroque ecology." "Baroque or complex ecology," she writes, suggests

> the spatial formation that repositions the city-state as a hub in an ecosystem created from the mobilization of diverse global elements—knowledge, practices, and actors—interacting at a high level of performance . . . The Singaporean ecosystem puts into play multiple domains, technologies, and actors so that their futures become intertwined in a network of sustainability. (Ong 2007: 180)

This is perhaps best exemplified by the concept of the growth triangle (GT), in effect a regionalization strategy meant to enhance regional advantage in the global economy (Ong 2007: 88). In Singapore's case, the GT strategy is pursued through the so-called Sijori, or Singapore-Johore-Riau GT, which connects the city-state to both Malaysia and Indonesia in an attempt to align each state's labor and technical resources in complimentary ways (Ong 2007: 89). From Singapore's perspective, the advantage of the GT is that it allows the city-state "to retain command/control functions at home while moving 'low-end' jobs offshore. It takes advantage of cheap Indonesian labor, and it also ameliorates tensions over the presence of too many guest workers within the city-state" (Ong 2007: 89).

Baroque ecology, however, also has symbolic-cultural ramifications. This is the second arena in which the claims of the Singaporean state over territory and citizens are being subtly reformulated. The management and efficiency of the emergent ecology, writes Ong, require "a complex reorganization of ethical norms . . . the knowledge ecology that has been forged is meant to shake Singaporeans out of their conventional ambition of working for multinational corporations and instead become knowledgeable, risk-taking entrepreneurs who help attract investments from global firms" (Ong 2007: 180, 183).

"A kind of biosociality" has emerged in which Singaporeans often have to question deeply held beliefs to fully participate in the new neoliberal order (Ong 2007: 184). Examples include pharmaceutical-industry provocations of Malays to overcome reservations about surgery and organ transplants, and the problematization of Chinese values of *guanxi*, kin- and community-based definitions of economic competition. Thus, according to Ong, the ethical and symbolic implications of being Singaporean are being reshaped by environments dominated by global pharmaceutical firms and other multinational corporations in which the figure of the Western entrepreneur takes precedence over that of the "traditional" Chinese merchant (Ong 2007: 185–189).

In the case of Dubai, the neoliberal turn has, similarly, initiated a process of dialectical adaptation of values—the neoliberalization of selected local ideas about capitalism and entrepreneurialism, but also the Arabization of neoliberal values—as part of a larger strategy, similar to that of Singapore's, to make the Emirati city-state a knowledge economy hub (Kanna 2010, 2011). Perhaps even more explicitly than is the case with Singapore, the ruler and the Executive Council (as already mentioned) analogize the city as a corporation. The ruler calls himself the "CEO of Dubai, Inc." Gently critiquing this development, an influential Emirati academic has deployed the terms *al-madina al-marka* (the city as brand), *al-madina al-qudwa* (the Dubai model), and *al-madina al-sharika* (the city-corporation) to capture the emergent Dubai reality (Abdulla 2006). Summarizing Dubai's Arabized version of neoliberalism entails:

> Non-interference by the state in the economy is the law for all aspects of life in the city-corporation . . . a city which offers a pool of skilled, cheap labor, 100 percent profit repatriation, non-interference in any way by the UAE state in the operations of the free zones, especially when it comes to labor laws, as well as lack of taxes and speed and ease of communications with other cities and capitals of trade and finance. (Abdulla 2006: 15)

Moreover, as in Singapore, the neoliberalization of Dubai has entailed a certain targeting of Dubai citizens in the realms of their bodily being and subjectivity, a kind of biopolitics carrying a set of technologies of the self-aimed at reshaping this subjectivity.[6] An example of this can be seen in the governmentality (Foucault 1991) of the knowledge economy enclaves of Dubai (Kanna 2010, 2011). These are free zones such as the Knowledge Village, Dubai Media City, and Dubai Internet City—known collectively as the Technology and Media Free Zone (Tecom)—where the state provides the infrastructure and exceptional governance regime to attract multinationals to Dubai. To manage Tecom, the state puts in place a member of the Executive Council at the head of the organization (during the time of my fieldwork, this was Ahmad bin Bayyat) and recruits young, English-speaking, often Western-educated Emiratis to positions of managerial influence. These Tecom employees regard both the organization and themselves as exceptional, well-educated, and uniquely prepared—in terms of social background and temperament—to embody and articulate what they call the "vision" of the ruler and the Executive Council. This vision, predictably, is expressed well by such figures as Thomas Friedman: a Dubai that has supposedly disengaged itself from what is stereotyped as an Arab world of tyranny, state socialism, and anti-Western attitudes, and which has "properly" adapted the lessons of a West valorized as entrepreneurial, progressive, and selfless in its dealings with the non-Western world (Kanna 2007, 2010, 2011).

Like their Singaporean counterparts, an individual from among the elite class in the Dubai population is selected by the neoliberal state and provided with an exceptional regime of rights and obligations. In turn, by internalizing and embodying this regime, these subjects indicate their capacity to take up the vocation of being a "worthy citizen," as Ong has put it in the context of Singapore. By doing so, these subjects enter into a "moral calculus" which requires them "both to excel at self-management and to be globally competitive and politically compliant" (Ong 2007: 194).

CONCLUDING THOUGHTS

Before concluding, it is important to mention that whereas Dubai and Singapore's trajectories have many similarities, there are also important differences.[7] Unlike Singapore, for example, Dubai's building spree was largely based upon debt, with EMAAR's signature project, the Burj Dubai/Khalifa skyscraper, costing approximately $1 billion and other typical projects exceeding this. Singapore's main government-owned holding company, Temasek, did not undertake projects like the Burj Dubai and other leveraged Dubai initiatives, focusing instead on infrastructural investment and basic services. In 2009, the major Dubai parastatal Dubai World, by contrast, could count 60 percent of its over $99 billion in assets as leverage. Moreover, unlike Singapore, where manufacturing accounts for a fifth of the economy, Dubai has an insignificant manufacturing sector. Virtually all of the sectors into which Dubai invested over roughly the past decade were related to services: property, tourism, and financial services, for example. These differences afford Singapore some protection against global market cycles, unlike Dubai's much more exposed economy.

Yet, for two city-states so geographically remote and culturally different from each other, Dubai and Singapore have also exhibited striking similarities in their trajectories. One reason, surely, is the geographic factor. Although Singapore long preceded Dubai as a major Asian entrepôt—it achieved this status soon after coming under British control in the first half of the nineteenth century, over a century before Dubai would become important in Asian/Indian Ocean commodity and capital flows—that they both become entrepôts is largely owing to the fact that they are both so-called chokepoints on important sea routes and, specifically, lie along the circuits of an Indian Ocean maritime political economy. Unlike other Asian developmental-capitalist states, such as Japan, South Korea, and Taiwan (with which both Dubai and Singapore have some similarities), the two Asian city-states exploited their geographic locations by engaging in entrepôt development. Lacking the agricultural hinterlands available to the larger Asian territorial states, both Singapore and Dubai eschewed territorial-national development and acted as intermediaries in Asian and global trade networks (Hvidt 2009: 400).

Moreover, Dubai's governance borrows from the model of Singapore, and actually seems to take it to more extreme levels. Singapore is of course governed by Lee Kwan Yew's PAP. In the PAP is concentrated enormous, unaccountable power legitimized by an ideology of merit in which PAP members are represented as the "best and the brightest" of Singapore. At (the largely nominal) elections, PAP candidates' résumés and educational certifications are openly displayed, asserting a right to govern by virtue of their allegedly superior intellect (Englehart 2000). In Dubai, power is even more concentrated. Even nominal elections are nonexistent. Five men—the ruler Muhammad Al Maktoum, his brother Hamdan, along with EMAAR chairman Muhammad Al Abbar, Dubai Holding chair Muhammad Al Gergawi, and Dubai World chair Sultan Ahmad bin Sulaym—control the economy (Hvidt 2009). In both Singapore and Dubai, therefore, the state is characterized by its dirigisme: decisions are made exclusively at the top, with citizens and other urban residents expected to go along with these without debate (Hvidt 2009: 401–402; see also Kanna 2011 for a more detailed discussion).

This type of governance has roots in the British imperial encounter. The cultural and ideological orientations both of upper PAP MPs and of the Al Maktoum are British: Lee Kuan Yew, his son and current Singapore Prime Minister Lee Hsien Loong, along with foreign minister George Yeo, were educated at Cambridge; S. Rajaratnam, a towering figure of Singapore politics in the second half of the twentieth century, was educated at the Raffles Institution in Singapore and, until the World War II interrupted his education, King's College London. Dubai's Rashid bin Said Al Maktoum, meanwhile, sent his sons to the Military Academy at Sandhurst. His son Maktoum, the current ruler Muhammad's late older brother, who ruled Dubai between 1990 and 2006, was well-known for his feudal English affectations and love of English country life.

This ideological formation, I suggest, had a deep impact on the world-view of the Singapore and Dubai rulers, one in which elites with a Western orientation and who can flexibly articulate themselves in both local and Western (and specifically upper-class English) cultural and linguistic idioms are seen to be endowed with the ability to know and best express the interests of their subjects. Both ruling classes have, therefore, actively repressed and marginalized other voices, other interpretations of modernity, in their societies. The Al Maktoum and their allies cast aspersions such as "Nasserist" or "agent provocateur" with as little discrimination as was the term "communist" in Singapore. The internal security apparatuses of each state took on immense power in the postindependence periods. These are all the legacy of the British tutelage of these regimes, in which indigenous people were routinely regarded with apprehension and targeted as objects of centralizing state discipline, and in which pliable local actors were seen as "moderate" and "modernizing" allies.

Let me conclude with a theoretical speculation. The argument about capitalism that is current in anthropology is that capitalism does not expand in a homogeneous way, overpowering the world with one model of the market economy. As geographer David Harvey tells us, capitalist expansion works much better if it accommodates cultural heterogeneity. Anthropologists such as Aihwa Ong and others (Hoffman et al. 2006) have extrapolated this insight more directly into the cultural sphere, and Harvey's influence on urban studies is also well-known. I owe lots to this work, and my own work has benefited in particular from Ong's insights into the interactions between processes of political economy and culture in Southeast Asia. The main thrust of this work is that capitalist globalization does not homogenize; quite the contrary: heterogeneity—geographic, functional, and cultural—is optimal.

The tradition of critical scholarship to which this literature is responding is captured well by the Marxist historian William Appleman Williams, who wrote of Adam Smith: "His entire system was predicated upon unending growth—upon empire . . . His objective was the same [as that of the mercantilists who he was critiquing], and he described it with great verve: 'the prosperity . . . the splendour, and . . . the duration of Empire' " (Williams 2007/1980: 77). Or take Fernand Braudel, not a Marxist, but here in agreement with the Marxist traditions: "[Capitalism] merely requires a way in, a foreign but colluding social hierarchy which extends and facilitates its action . . . the connection is made, the current transmitted" (Braudel 1986: 65). The sense here is not one of heterogeneity, primarily, but about a single underlying reality: empire and expansion of markets. Heterogeneity may be the apparent reality, but homogeneity is the undercurrent. To adapt a metaphor from Braudel: empire is the ocean current upon which develops the foam of culture, society, and politics. My main question for future research is this: should the neoliberal trajectories of Singapore and Dubai give us pause to rethink the more recent critiques of capitalism, with their emphasis on heterogeneity? The striking echoes of the image of Curzon's expedition to the Trucial Coast over a century ago in that recent *New York Times* ad is certainly provocative in this regard. It suggests that, at least in some sectors of the global economy—perhaps its more "successful" sectors?—the line between imperial capitalism and neoliberal capitalism is a direct one.

NOTES

1. See http://www.emaar.com/index.aspx?page=home. Accessed January 15, 2012.
2. More specifically, Dubai's International Airport, DXB, is modeled on Singapore's Changi Airport, and Dubai's Emirates Air explicitly seeks to rival Singapore Air. Dubai Ports also sees PSA, the Singapore Ports Authority, as its model and competitor (Oh 2009).
3. Christopher Parker and Pete Moore write about "southern (Iraqi) cities overrun with goods coming over the border from Iran and re-exported from Gulf ports, primarily Dubai" (Parker and Moore 2007: 13).
4. Fenelon, for example, tells us that the carrying trade was a central Dubai economic activity throughout the nineteenth century. Between December and March, sailing vessels made use of the northeast monsoon to take them

to Mombasa, Zanzibar and Dar Es-Salaam, whereas between April and September, they returned with the southwest monsoon (Fenelon 1976: 64).
5. All figures in this paragraph, unless otherwise noted, are taken from Marchal 2005, 103–104.
6. My thanks to Farha Ghannam for pointing this out in her generous, critical, and thoughtful reading of an earlier essay (Kanna 2009).
7. Data and figures in this paragraph are based on Oh 2009.

REFERENCES

Abdulla, Abdel Khaleq (1984). *Political Dependency: The Case of the United Arab Emirates*. Ph.D. Dissertation. Georgetown University Department of Politics.

Abdulla, Abdel Khaliq (2006). "Dubai: rihlat madina 'arabiyya min al-mahalliyya ila l-'alamiyya" ("Dubai: The Journey of an Arab City from Regionalism to Cosmopolitanism"). *Al Mustaqbal al-Arabi* 323 (January): 1–28.

Adelkhah, Fariba (2001). "Dubaï, Capitale économique de l'Iran?" Pp. 39–65 in *Dubaï: Cité Global*, edited by Roland Marchal. Paris: CNRS.

Al-Sayegh, Fatma (1998). "Merchants' Role in a Changing Society: The Case of Dubai 1900–1990." *Middle Eastern Studies* 34 (1): 87–102.

Bose, Sugata (2006). *A Hundred Horizons: The Indian Ocean in the Age of Global Empire*. Cambridge, MA: Harvard University Press.

Bowcott, Owen (2002). "Criminal Trail That Leads from Carjackings to Streets of Asia." *The Guardian*, February 22.

Braudel, Fernand (1986). *The Perspective of the World: Civilization and Capitalism 15th–18th Century*, vol. 3. Sian Reynolds, trans. New York: Perennial.

Daoud (2001). "US Targets Bin Laden's Money Men." *The Guardian*, November 8.

Chee, Soon Juan (2001). "Pressing for Openness in Singapore." *Journal of Democracy* 12 (2): 157–167.

Chew, Ernest C. T. (1991). "The Foundation of a British Settlement." In *A History of Singapore*, edited by Ernest C. T. Chew and Edwin Lee. Singapore: Oxford University Press.

Davidson, Christopher M. (2008). *Dubai: The Vulnerability of Success*. New York: Columbia University Press.

Economist (2002). "Beyond Oil." *The Economist*, March 23: 26–28.

Englehart, Neal A. (2000). "Rights and Culture in the Asian Values Argument: The Rise and Fall of Confucian Ethics in Singapore." *Human Rights Quarterly* 22 (2): 548–568.

Fenelon, K. G. (1976). *The United Arab Emirates: An Economic and Social Survey*. New York: Longman.

Foucault, Michel (1991). "Governmentality." Pp. 87–104 in *The Foucault Effect: Studies in Governmentality*, edited by Graham Burchell, Colin Gordon, and Peter Miller. Chicago: University of Chicago Press.

Ho, Engseng (2002). "Gangsters into Gentlemen: The Breakup of Multiethnic Conglomerates and the Rise of a Straits Chinese Identity." Paper presented at The Penang Story International Conference, Penang, Malaysia, April 18–21.

Ho, K. C. (2009). "Competitive Urban Economic Policies in Global Cities: Shanghai through the Lens of Singapore." Pp. 73–91 in *Shanghai Rising: State Power and Local Transformations in a Global Megacity*, edited by Xiangming Chen. Minneapolis: University of Minnesota Press.

Hoffman, Lisa, Monica DeHart, and Stephen J. Collier (2006). "Notes on the Anthropology of Neoliberalism." *Anthropology News* 47 (6): 9–10.

Hvidt, Martin (2009). "The Dubai Model: An Outline of Key Development-Process Elements in Dubai." *International Journal of Middle East Studies* 41: 397–418.

International Oil Daily (2007). "Nigeria Starcrest Faces Probe." February 13.

Janardhan, Meena (2007). "Gulf States Fight a Flood of Dirty Money." Inter Press Service, March 19.

Kanna, Ahmed (2007). "Dubai in a Jagged World." *Middle East Report* 243 (Summer): 22–29.

Kanna, Ahmed (2009). "Flexible Citizenship and Ambivalent Arab Identity in the Arab Gulf." Invited Paper, NYU Center for Near Eastern Studies Colloquium Series, 26 January.

Kanna, Ahmed (2010). "Flexible Citizenship in Dubai: Neoliberal Subjectivity in the Emerging City-Corporation." *Cultural Anthropology* 25 (1): 100–129.

Kanna, Ahmed (2011). *Dubai, the City as Corporation.* Minneapolis: University of Minnesota Press.

Ken, Wong Lin (1991). "The Strategic Significance of Singapore in Modern History." In *A History of Singapore*, edited by Ernest C. T. Chew and Edwin Lee. Singapore: Oxford University Press.

Khaleej Times (2005). "Unpaid Workers Are the New Newsmakers." September 26.

Longva, Anh Nga. 2005. "Neither Autocracy nor Democracy but Ethnocracy: Citizens, Expatriates and the Socio-Political System in Kuwait." In *Monarchies and Nations: Globalization and Identity in the Arab States of the Gulf*, ed. Paul Dresch and James Piscatori, 114—135. London: I.B. Tauris.

Marchal, Roland (2005). "Dubai: Global City and Transnational Hub." Pp. 93–110 in *Transnational Connections and the Arab Gulf*, edited by Madawi Al Rasheed. New York: Routledge.

Oh, Boon Ping (2009). "Singapore and Dubai: Alike, Yet so Different." *Asia One*, December 6. Electronic document, http://www.asiaone.com/Business/News/Story/A1Story20091204-183900.html, accessed August 26, 2010.

Ong, Aihwa (2007). *Neoliberalism as Exception: Mutations of Citizenship and Sovereignty.* Durham, NC: Duke University Press.

Onley, James (2005). "Britain's Informal Empire in the Gulf, 1820–1971." *Journal of Social Affairs* 22 (87): 29–45.

Parker, Christopher, and Pete W. Moore (2007). "The War Economy of Iraq." *Middle East Report* 243 (Summer): 6–15.

Soon, Lau Teik (1991). "Singapore in South-East Asia." In *A History of Singapore*, edited by Ernest C. T. Chew and Edwin Lee. Singapore: Oxford University Press.

Starostin, Dmitry (2007). "Dubai Gets Russified." *Moscow News*, July 5. Electronic document, http://www.mnweekly.ru/feature/20070705/55260665.html, accessed October 31, 2007.

Thio, Eunice (1991). "The Syonan Years, 1942–1945." In *A History of Singapore*, edited by Ernest C. T. Chew and Edwin Lee. Singapore: Oxford University Press.

Williams, William Appleman (2007/1980). *Empire as a Way of Life.* Brooklyn, NY: Ig Publishing.

Zahlan, Rosemarie Said (1978). *The Origins of the United Arab Emirates: A Political and Social History of the Trucial States.* London: Macmillan.

3 Shaping Politics in Chinatown

The Intersection of Global Politics and Community Politics in Wartime and Cold War San Francisco

Scott H. Tang

Those exploring San Francisco's St. Mary's Square for the first time will no doubt notice the towering statue of Sun Yat-sen. Twelve feet of stainless steel with red granite for Sun's head and hands, the striking figure of the Chinese revolutionary and the leader of the Republic of China draws everyone's attention. Its presence in the space next to Old St. Mary's Cathedral is hardly an anomaly because the Chinatown that was rebuilt after the 1906 earthquake and fire eventually expanded along the streets adjacent to the square. In other words, the statue itself may be remarkable, but its location in Chinatown is not. To viewers, it appears to be one of many signs of ethnic difference in the urban landscape and a confirmation of the city's famed ethos of tolerance.

However, applying today's standards of cultural pluralism to explain the statue ignores its historical and political significance. Erected in 1937, the statue made tangible a gradual transformation in how white San Franciscans perceived their Chinese neighbors. William Lawson, a California administrator for the Work Progress Administration, proclaimed at its dedication:

> We, of San Francisco, enjoy the closest association with our own admirable Chinese population. Kindly, generous, honorable, steadfast in friendship they have earned our very great respect. The Chinese stock in San Francisco is pioneer stock. The Chinese here are largely native born. Their grandfathers and great grandfathers were brave, adventurous souls. They cut their bridges behind them, crossed the wide Pacific and made their way and their peace in a strange civilization among people whose language they did not speak and whose ways seemed the ways of barbarians to those whose background lay in the culture of the Orient. They met the West and today their descendants are our friends and our esteemed fellow citizens. The Chinese are admirable business men. Their stores, with their wares so strange to the people of our country are a great attraction in our city. We prize them very highly.[1]

As Lawson uttered words of praise that seldom surfaced in San Francisco half a century earlier, Dennis Kearny and other proponents of Chinese exclusion must have been rolling in their graves. Chinese un-assimilability had been one of their main arguments for exclusion, and Chinatown itself was evidence enough of the immigrants' refusal to cut their bridges.

Lawson's speech did more than conceal an earlier history of white Americans disparaging Chinese immigration and mobilizing to curtail it. The speech obscured the fact that Chinese San Franciscans still experienced immigration restriction laws and still encountered difficulties in attaining decent jobs and homes. Despite these past and present realities, Lawson's words established that residents of Chinese descent should be valued for their contributions to the city and treated as fellow San Franciscans. Chinese immigrants met the West and became Westerners themselves.

Most importantly, the story behind the statue of Sun Yat-sen provides insight on the intersection of politics in China and life in Chinatown. Like many immigrant groups, Chinese immigrants remained interested in developments in their homeland. While visiting overseas Chinese communities to raise funds for his revolutionary movement, Sun found immigrants who were eager to play a role in overthrowing the Qing Dynasty and improving life in China. Some immigrants contributed by rallying support or working for *Young China*, the Revolutionary Party's newspaper. Others tried to help more directly by returning to Asia (Nee and Nee 1973: 74–75). Moreover, when they later commissioned the statue to commemorate Sun's visit, Chinatown leaders were keenly aware of the implications for U.S.-China relations during the Second Sino-Japanese War. Lionizing Sun, who symbolized the Chinese adoption of modern values and republican principles, bolstered American sympathy for China as the nation Sun helped create struggled against Japanese militarism.

Those involved in the China-focused politics also hoped their actions would improve life in Chinatown. Leland Chin recalled years later that feelings of alienation in the immigrant community, particularly among the laboring classes, made Sun's message attractive:

> They've been discriminated against, and pushed around so much, I mean our people in this country had been pushed around so much they wanted some way, sort of revenge, to get themselves a better way to survive. . . . You can't even go into a restaurant and sit down and enjoy a meal. You're afraid to go in the barbershop and get yourself a haircut. (Nee and Nee 1973: 78–79)

In their minds, strengthening China and earning the respect of the West could lead to better treatment in America (Nee and Nee 1973: 78–89). Similarly, Chinese Americans in the 1930s tried to use their association with China to improve relations with white Americans. Lawson's references to exotic goods and admirable businessmen suggest the importance of Chinatown tourism. After reminding readers that money spent in Chinatown added to the China war relief fund, a brief note in *Time* magazine characterized the statue of

Sun as a joint Chinese and American effort: "Materials were provided by the Kuomintang (Chinese Nationalist Party) in the U. S. Labor was WPA" ("Art: Statues" 1937: 35). Along with other attempts to showcase Chinese assimilability, Chinatown tourism, and Chinese American cooperation, the statue of Sun Yat-sen cast all Chinese in a positive light.

This chapter examines politics in San Francisco's Chinatown in the middle of the twentieth century to understand not only the development of this ethnic community's political culture, but also the relationship between activities on the local level and the world outside America's national borders. The selected time frame includes two international crises involving the United States and China—the Second World War and the Cold War. Throughout the period, Chinese Americans highlighted their association with China and acted to help their ancestral land and secure their positions in America. Even the end of the Kuomintang hegemony in Chinatown grew out of inseparable global and local causes. This confirms Michel Laguerre's argument that one could consider the local and the global as a connected reality rather than two discrete realities (Laguerre 1999: 37–38). The global does not simply shape the local or vice versa. Diasporic politics such as those in Chinatown have intertwined local and global roots and simultaneous local and global implications (see Moser, this volume, on the simultaneity of both in very different transnational contexts).

Recognizing this connected reality moves us away from earlier nation-based paradigms in immigration studies that focused on the push and pull factors behind immigration, the severing of Old World ties, and the assimilation and acculturation processes. We instead see ethnic Americans operating within transnational networks that play a significant role in an ethnic community's economic, social, political, and cultural development. This historical analysis of Chinatown exposes political dimensions that can exist in almost any ethnic community. Like the Irish Americans involved in the nineteenth-century Fenian movement and the hard-line Cuban Americans who continue to resist rapprochement with the Cuban government, Chinese Americans fashioned their ethnic and political identities in response to the interplay of foreign and domestic concerns. The Chinatown case study also reveals how these identities found expression in urban cultural production, such as the ethnic press, the annual Double Ten celebration, and the Sun Yat-sen statue (see Figure 3.1).

Chinatown's association with China, which ethnic leaders could leverage for their benefit in the twentieth century, was an important component of the nineteenth-century anti-Chinese movement. Besides blaming Chinese immigrant labor for white unemployment and wage degradation, critics at that time decried the transplanting of Chinese culture onto American soil. Chinatown stood as a constant reminder to white San Franciscans that "peculiar" aliens lived among them, representing a failure to assimilate that endangered the goal of a homogeneous community (see Notar, this volume, on how residents of a changing and less homogeneous city in

China view outsiders). Those who opposed immigration made Chinatown into the source of all urban evils, including opium, prostitution, and gambling, and no description of Chinatown seemed complete without some reference to odors, overcrowding, or vermin. The critique of the Chinese presence also included the positioning of the immigrants as a danger to white political dominance and American political institutions. As early as

Figure 3.1 Statue of Dr. Sun Yat-sen in Chinatown, 1937. San Francisco History Center, San Francisco Public Library.

1854, California Supreme Chief Justice Hugh Murray warned in a majority opinion that if the state recognized the right of Chinese immigrants to testify against whites, citizenship rights would follow, and "we might soon see them at the polls, in the jury box, upon the bench, and in our legislative halls" (Ringer 1983: 582583). Reverend Milton Starr later argued that the Chinese were a servile race that would "vote against the short-haired Caucasians." He described to audiences the following scenario:

> Suppose any great monopoly, or that the monopolists united, should desire to carry an election, and fix a legislature to pass any subsidy law, they would only have to buy the hundred thousand votes from the six Chinese companies who, with long nails and whip-thong tails, sit in judgment, like princes over a dominion of voting slaves, and make them do their will. (Starr 1873: 102–103)

This representation and analogous rhetoric concerning "despotic mentality" justified political marginalization. In the end, government authorities asserted that the 1870 naturalization act barred Chinese immigrants from becoming citizens and thereby made these nonwhites permanent aliens with virtually no political power.

Additional circumstances surrounding Chinese immigration shaped political culture within Chinatown. The immigration that occurred in the second half of the nineteenth century was primarily men seeking overseas economic opportunities to support families in China. Many of them saw themselves as sojourners, even if permanent overseas residence was the reality of their lives. Chinatown was a world of men separated from their families, and the gender imbalance, which was reinforced by immigration restriction, delayed natural growth. As a result, American-born Chinese males, who would have voting rights once they reached adulthood, were an insignificant number before the twentieth century.

Denied citizenship, the immigrant generation could not fully participate in American politics and could not rely on government authorities to address their needs. Living in an enclave offered some protection, but only further isolated the immigrants from mainstream society. Moreover, Chinese Americans customarily eschewed personal involvement in the political world outside of Chinatown because many of the immigrants had circumvented the racist exclusion laws to enter the country illegally. Rather than draw attention to themselves, Chinese Americans allowed the Chinese Consolidated Benevolent Association (CCBA), a federation of regional association leaders that was more commonly known as the Chinese Six Companies, to speak for them in all interactions with external political entities. Chinese immigrants continued to participate in ethnic organizations including surname and regional associations, both of which offered essential services in a comfortable Chinese-language environment. These mutual-aid organizations maintained rooming houses for temporary lodging, connected their members with employment agents, and assisted with burial services (Ma 1991: 97–98, 150).

At first, regional association presidents, and by extension the leaders of the CCBA, were all gentry-scholars who had passed civil service examinations in China and subsequently migrated to take their posts in America. This system of appointments, which was in place until the 1920s, resulted in the preservation of Chinese political traditions as well as the close relationships between Chinatown's political leaders and Chinese foreign diplomats such as the Chinese consul general in San Francisco. Later on, the CCBA would become dominated by local merchants, which ultimately led to the organization's class orientation (Chen 2000: 110–114, 319).

The political dominance of the immigrant generation and their organizations' transnational connections reinforced the attention paid to developments in China. Chinese Americans followed overseas events with great interest, especially because the immigrants still had family members in China. This explains why Chinatown's residents raised relief funds following natural disasters in the lower Yangtze River area and enthusiastically embraced the nationalist movement (Yung 1995: 98). All the while, Chinese-language newspapers, several of which were organs of political parties in China, kept everyone China-focused. According to James Low, "If you were Chinese-American you certainly felt the fate of China was important. In fact we followed Washington politics very little!" (Nee and Nee 1973: 170).

Over time, the Kuomintang (KMT) established a strong presence in Chinese American communities. In the 1910s, members of the Chinese political party played a leading role in mobilizing communities to boycott Japanese goods and services and participated in the formation of the China Mail Steamship Company as an alternative to using Japanese ships. The party would later use vernacular dramas and Chinese-language school lessons to promote its nationalistic doctrines. These activities were meant to arouse Chinese nationalism and to generate party support among overseas Chinese communities. The KMT's reliance on these communities was substantial; as late as 1929, over 25 percent of the party's members were overseas Chinese. By the end of the next decade, the party had successfully recruited Chinatown leaders to join its ranks and permitted them to rise to high positions within the American branches of the party. The KMT political network became even stronger with the formation of the China War Relief Association in 1937. Chaired by the president of the CCBA, the association coordinated the relief efforts in Chinatown and in Chinese communities throughout the United States. The overlapping leadership of the KMT party in America, the CCBA, and the Chinese War Relief Association positioned the KMT as the dominant political force locally, nationally, and transnationally (Lai 1991: 183, 194–197).

Whereas the CCBA was the leading political institution in the period before the war, organizations that focused on issues of American citizenship were also part of Chinatown's political world. The Chinese American

Citizens Alliance (CACA) emerged to teach the growing American-born generation about their rights and responsibilities as American citizens. It stressed loyalty to the United States and argued that pledging allegiance to China endangered American citizenship and the reputation of those trying to establish their homes in America. The group also served as a mutual-benefit association, providing financial assistance and life insurance to its members, and published a Chinese-language newspaper (Chen 2000: 208–212).

Although it resembled an exclusive fraternal organization and had its fair share of Chinatown elites, the CACA encouraged broad political involvement and engaged in American-style political activities (Chung 1998: 106). During election season, it posted public statements with each candidate's position, assigned representatives to field questions concerning the election, and established classes to teach eligible voters how to register and vote. Alliance members argued that mobilizing voters would bring the permanent improvements that the CCBA and the Chinese government seemed unable to achieve. As early as 1917, an editorial in *Young China* told readers that the "Zhonghua Huigan [CCBA] can only provide some temporary remedies but cannot take out the firewood from under the cauldron and stop the water from boiling" (Chen 2000: 211). However, minor criticism of the CCBA did not prevent cooperation among Chinatown organizations. The large number of professionals and businessmen in the CACA ensured that the organization and the CCBA had enough shared interests and orientations to work together effectively.

Despite the early emphasis on voting as the only way to improve their condition, many Chinese American citizens revealed indifference toward American politics. The high levels of apathy stemmed from feelings of alienation and the general belief that minority political participation was meaningless. One American-born Chinese described his right to vote as "an empty gesture" and believed that he would remain forever a second-class citizen in his native land (Dobie 1936: 325). Lim Lee, who later helped found the Chinese American Democratic Club and served as postmaster general of San Francisco, noted in 1940 that only 50 percent of the seven thousand eligible Chinese American voters chose to register and criticized apathetic citizens for allowing the immigrant generation to control the outcome of local campaigns. "It is justly our fault that politicians cooperated with alien demagogues to run our politics for us," Lee wrote (Lee 1940: 1).

Feeling marginalized in the American political system did not necessarily mean that Chinese Americans were politically apathetic. As involvement in both revolutionary and reform movements showed, many Chinese Americans were interested in politics and hoped a better China would improve the lives of Chinese throughout the world. According to Sue Fawn Chung, the rising political consciousness concerning China led to interest in Chinese American rights in the United States and the expansion of the CACA into a national organization (Chung 1998: 101).

The conflict between China and Japan provided another example of how global politics and local concerns were intertwined. On September 24, 1931, less than a week after Japanese troops began occupying Mukden (today's Shenyang, Liaoning province), a city in Manchuria, the CCBA and other Chinatown groups formed the Anti-Japanese Chinese Salvation Society. One of the society's objectives was to organize a boycott of Japanese products (Yung 1995: 224–225). Ethnic competition also came to be viewed through the lens of war. At the end of the decade, Chinese newspapers warned its readers that Japanese Americans were occupying commercial real estate along Grant Avenue, one of Chinatown's main thoroughfares. They described the phenomenon as an invasion of Chinese territory and drew comparisons with what was happening abroad (Lowe 1936: 90). In this period of bolstered ethnic identification, the common problems facing Asian Americans—e.g., employment discrimination, residential segregation, and the denial of immigration and naturalization rights—never became the most important issues to be addressed.

The overseas conflict became a part of daily life in 1930s Chinatown as men, women, and children mobilized to raise money and awareness. Pardee Lowe characterized the interest in the war as "passionate, almost fanatical" and gave the following description of the community at that time:

> Banners straddle streets announcing theatrical benefits and war drives. Sound trucks blare forth the latest news in Cantonese. Little moppets shine shoes and sell magazines for war relief. Extras of the community's five language newspapers are available at every corner fruit-vendor's stand. In the book stores are huge lithographs bearing the idealized countenances of Generalissimo Chiang Kai-shek and his Wellesley-educated wife and countless books on the Sino-Japanese undeclared war. In every show-window hang huge posters in English urging American friends to refrain from patronizing Japanese stores while in Chinatown or urging a general boycott of Japanese goods.[2]

A particularly dramatic demonstration occurred on December 16, 1938, when Chinese San Franciscans and sympathetic whites protested the loading of scrap iron onto a ship bound for Japan. Organized by the Chinese Workers Mutual Aid Association, the protest began with over five hundred demonstrators, three-fifths of whom were white, picketing the *S.S. Spyros* on Pier 45. The mass demonstration lasted four days, and the number of protesters grew to an estimated five thousand people. Those manning the picket line included longshoremen who had walked off the job in protest and Chinese American men and women of all ages, classes, and political affiliations. Chinatown restaurants and grocery stores showed their support by giving free food and drinks to the protesters. Even though the scrap iron made its way onto the ships after the demonstration had been called off, the mass protest made the public more aware of America's role in the

Sino-Japanese War and led to renewed calls for an embargo on all materials to Japan (Yung 1995: 241–243).

Full participation in the war effort encouraged Chinese American women to expand their public role through the women's auxiliary to the Chinese War Relief Association and mainstream American organizations such as the YWCA. For example, the women involved in the Chinese YWCA voiced their ethnic community's interests by urging all women to boycott Japanese silk. "If every American girl would refuse to wear silk stockings, Japan's invasion of China would stop overnight," a newsletter stated. "Every girl who delays joining the movement is unwittingly helping to postpone the ending of Japanese aggression."[3] The general political atmosphere also influenced the youngest residents of Chinatown. Lorena How, whose mother engaged in boycotts and fund-raisers, told an interviewer that as a young child she used to run down to the Japanese-owned stores on Grant Avenue to drive away white customers. She would yell in English, "*Don't buy, lady, Japanese store!*" until broom-wielding Japanese ladies chased her away. How's commitment to boycotting Japanese goods was so strong that she even returned a doll that had been given to her simply because it had been made in Japan (Yung 1999: 506).

During the 1930s, being identified with China actually improved how white Americans saw Chinese San Franciscans. Positive images of the Chinese, especially those found in Pearl Buck's *The Good Earth* and Henry Luce's magazines, laid the groundwork for this change in white racial attitudes. Instead of being peculiar and inferior, the Chinese became farmers struggling to keep their land in the face of natural disaster and social calamity. Articles in *Time*, *Fortune*, and *Life* added to the makeover with favorable descriptions of social and political reformation in China and an unwavering praise for China's leaders. They made Chiang Kai-shek into a savior, the one who would Americanize China by bringing Christian salvation, political democracy, and economic modernization. *Time* even named Chiang Kai-shek and Soong Meiling (Madame Chiang Kai-shek) their man and wife of the year in 1937. These representations in American culture translated into white sympathy for China after the outbreak of the Sino-Japanese War made the Chinese people the victims of Japanese military aggression. At the end of the decade, 74 percent of Americans expressed support for China in the Sino-Japanese War whereas only 2 percent expressed pro-Japan sentiment. The U.S. government also became more involved in the war by offering Chiang lend-lease aid in November of 1940 (Jespersen 1996: 25–26, 39–44; Chun 1993: 75).

In San Francisco, Chinese Americans engaged in their own activities to subvert the negative images of Chinatown. Participation in the 1939 Golden Gate International Exposition provided the opportunity to introduce to the white public a modern yet ethnically distinct Chinese immigrant community. The fair's Chinese village, which was designed by a committee composed of American-born Chinese and foreign-born Chinese merchants, allowed the approximately one million visitors to see the handiwork of Chinese artisans, to enter a Chinese temple, and to enjoy a Chinese garden. Exotic architecture such as the towering pagoda made Chinese culture appear magnificent. The attraction also had "The Good Earth Settlement," a constructed farming village that brought to life the world described in Buck's popular novel (Chun 1998: 180–184).

The cultural experience dispensed at the fair led to a tourism revival in Chinatown as whites began to consider the ethnic community to be a safe environment and to appreciate what it had to offer. This realization that Chinatown was a valuable tourist destination clearly added to the more sympathetic attitudes that whites had toward the Chinese. An article published by the Down Town Association of San Francisco in June of 1938 reminded readers: "Let us not underestimate its value to our city in dollars and cents. Chinatown is our magnet with power to attract tourists from all over the world." The author went onto recommend the "complete orientalization" of Chinatown with the introduction of lanterns, rickshaws, and Chinese girls clothed in colorful silks.[4] Pardee Lowe confirmed that "orientalization" was occurring in Chinatown to attract tourists. Among the several developments that he noticed was the establishment of modern drinking establishments in the community. He wrote:

> The cocktail bars may be a new phenomenon in the local scene but they are part and parcel of the current policy of Chinafying Chinatown for Profit. In less than eighteen months they have altered the entire outlook of the community. With tremendous commercial success, it has discovered that the Chinatown Cocktail Bar Orientale is an adequate substitute for the exotic local color that disappeared with the fire. Today there are more than a score of such elaborately decorated establishments which [sic.] vending Western and Chinese drinks in a cavernous setting suffused with neon lights, pseudo Chinese art, and native hostesses, singers, jazzbands and "emcees".

During the late 1930s, therefore, white Americans celebrated and consumed racial difference in Chinatown. Both practices reinforced the belief among white San Franciscans that they and their city did not have a race problem.[5]

America's formal entry into the Second World War two years later enabled Chinese San Franciscans to use America's wartime alliance with China to improve their image and their position in America. White Americans continually saw Chinese Americans as China's overseas representatives. For example, Kenneth Bechtel, the president of Marinship, praised Chinese American industriousness and patriotism in a letter sent to Chiang Kai-shek. He wrote:

> We have learned that these Chinese-Americans are among the finest workmen. They are skillful, reliable—and inspired by a double

> allegiance. They know that every blow they strike in building these ships is a blow of freedom for the land of their fathers as well as for the land of their homes. (Yung 1999: 474)

Jade Snow Wong, meanwhile, captured how those in Chinatown made new lives for themselves in the shipyards. She noted the mixture of foreign-born and native-born Chinese Americans at Marinship, commenting that "some are university graduates, some are older Chinese who were perhaps printers or cooks, some are women who were sheltered housewives" (Yung 1999: 478). Whether they were in unskilled or skilled positions, Wong argued, these Chinese American workers contributed to the shipyard's success. She told readers of the company newsletter:

> The paper picker-uppers who keep the yard neat, the cook who prepares wholesome food, the burner who cuts with precision and patience, the draftsman who draws with care and accuracy, the timekeeper who records working hours, the boilermaker helper who fills the buckets with essential shifters' hardware, the girl who makes travel reservations—these Chinese at Marinship are each in his or her own way working out their answer to Japanese aggression: by producing ships which will mean their home land's liberation. (Yung 1999: 478)

Local discussions concerning the repeal of Chinese immigration restriction further confirm the significance of the U.S.-China wartime alliance. Dan Mah, a Chinese American delegate from the Miscellaneous Employees Union, submitted a prorepeal resolution to the San Francisco Labor Council, which was affiliated with the American Federation of Labor. "We have to make a factual demonstration of our sincerity," Mah declared, "not just give lip-service to China" ("Labor Council Backs" 1943: 4). Vice President Dan Haggerty and Secretary John O'Connell, however, opposed the resolution and even revived the argument that Chinese immigration was a threat to wages. "A Chinaman is a Chinaman wherever he is," O'Connell asserted. Mah's resolution ultimately failed in a seventy-eight to fifty-four vote ("Labor Council Backs" 1943: 4).

The San Francisco Labor Council's position on the repeal differed from the one made by the Congress of Industrial Organizations. In April 26, 1943, the California CIO Council sent a statement backing the repeal drive to Madame Chiang Kai-shek and to Chinese diplomats in America. "Unity of the people of all the United Nations is vitally necessary to the winning of the present war against fascism," the CIO's statement read. "That unity must be not merely a matter of high-sounding statements but a real bond of sympathy between the peoples of the great democratic nations now engaged in struggle against the bandit nations of the Axis" ("CIO Hits Ban" 1943: 1).

When the San Francisco Board of Supervisors was considering a resolution to endorse pending repeal proposals, most of those who made oral presentations at the meeting expressed support for the repeal measure by stressing the current and future U.S.-China relationship. Three Chinese San Franciscans made their case for repeal by citing Chinese American contributions to the war effort. One speaker, who had spent thirty-eight years in China as a representative of the United States Chamber of Commerce, argued that improving ties with China would bring economic benefits. He said that San Francisco was the key city for export to China before speculating that relations with the Asian nation would increase productivity and trade to such a large degree that San Francisco would outshine present-day New York. After hearing the different opinions and delaying their decision until a city attorney confirmed that the revised policy would only benefit immigrants from China, the supervisors unanimously voted in favor of the prorepeal resolution (San Francisco [Calif.] Board of Supervisors 1943: 1964–1967; "Admit the Chinese!" 1943: 1).

The intersection of political developments abroad and local politics persisted in the postwar era, especially when the conflict between the KMT and the Chinese Communist Party resumed. A few left-leaning Chinatown organizations, including the Chinese Workers' Mutual Aid Association and the Chinese American Democratic Youth League, were vocal critics of both Chiang Kai-shek and the KMT, and they pledged their support for the revolutionary forces fighting for China's liberation. Chinatown's traditional political leadership, however, still maintained strong ties with the KMT. The CCBA continually stressed the legitimacy of Chiang's rule and sought to undermine Communist support in the community. These ideological and political differences brought China's political struggle to Chinatown.

The best example of this occurred right after the proclamation of the People's Republic of China in 1949. A joint committee of the Chinese Workers' Mutual Aid Association, the Chinese American Democratic Youth League, and a small group of liberal Chinatown businessmen decided to sponsor a celebration in the Chinese American Citizens Alliance hall. In addition to scheduled speeches, a choral group from the left-wing California Labor School was to perform as well. About eight hundred people, a few of them black and white members of the International Longshoremen's and Warehousemen's Union, filled the large auditorium on the night of the celebration. They listened to the speeches, applauding every time the PRC victory was mentioned. Everything was progressing as planned until a shout from the door interrupted the festivities. One witness recalled seeing two lines of men running through the aisles to the speakers' platform, tearing down the PRC flag, knocking over the vases of flowers, before throwing a blue dye at audience members. An account published in the *People's World*, the local Communist Party newspaper, reported that forty hoodlums, described as "mostly young men, some teen-agers, but led and directed by older mobsters," smashed chairs, threw eggs, and destroyed a movie projector. "The blackjack wielding thugs," as the paper called them, also slugged several spectators, causing head lacerations on two Chinese Americans. One white supporter received a broken rib while trying to protect a poster ("Hoodlums Fail" 1949: 2; Nee and Nee 1973: 214–215).

On the following night, Chinatown held its Double Ten celebration to commemorate the overthrow of the Manchu Dynasty. Community leaders fashioned the event to bolster the legitimacy of the KMT. The CCBA made the following statement: "The Chinese in America protest the Communist regime, re-pledge their loyalty to democratic China and reaffirm their faith in the principles of Dr. Sun Yat-Sen [see Figure 3.2], the George Washington of China." Speakers at the event likewise denounced the "Communist puppets of Peking," and solicited support and sympathy for Chiang's KMT forces ("Peking and the West" 1949: 1, 9). In addition, a leaflet printed in Chinese began circulating during the parade. It denounced several Chinese Americans as "Communist bandits" and called for their extermination. The leaflet's authors offered a 5,000-dollar reward for the death of anyone on the list. Those named, primarily newspaper editors and group leaders who either expressed sympathy toward the PRC or were merely critical of the Chiang regime, immediately went into hiding. This KMT-sponsored terrorism was only one of several efforts to silence the dissident voices of Chinatown leftists

Figure 3.2 Tse Kiong Sun placing flowers at the foot of statue dedicated to his grandfather Sun Yat-sen, 1943. San Francisco History Center, San Francisco Public Library.

("Chiang Stooges" 1949: 2; Kalman, "Protection Asked" 1949: 1; Kalman "Chiang Spies Here" 1949: 1; Nee and Nee 1973: 215).

However, leftists were not the only group in the Chinatown community that voiced anti-Chiang sentiment. Gilbert Woo, editor of the *Chinese Pacific Weekly*, estimated that in 1949 over 80 percent of the people in Chinatown believed that China's government under Chiang needed to change. Criticism of Chiang and the KMT "wasn't based on communism," Woo declared. "We just criticized the way we would criticize the president or the White House" (Nee and Nee 1973: 223–225). Meanwhile, to the surprise of one writer from the *Chinese Press*, some local importers believed that China under Communist rule would be good for business. "They think that a Red victory would bring peace, a stable currency and better trading conditions," the paper said ("China Imports Arrive" 1949: 1). Harry Chew of the Far East Importing Company made the same observation when a *Chronicle* reporter asked him about how Chinatown was responding to the civil war abroad. "I'm inclined to think most importers are very optimistic about the Communist setup," Chew conjectured before adding that trade ultimately would depend on the status of relations between the United States and China. "If the American Government says 'no soap, no ships to China' then we're out of business," he said.[6]

Tso Tang, the secretary of the Chinese Chamber of Commerce, summed up the general attitude of many Chinatown merchants. "With us, we are less concerned with political matters. It's matter of life and death, food and a living. We have to face reality, not theories. We are concerned with the economic life." His organization criticized KMT corruption and even passed a resolution calling for the extradition to China of T. V. Soong and K. H. Kung, two former prime ministers and KMT financiers. Some Chinatown residents believed the resolution was the merchants' attempt to ease future agreements with the Chinese Communists (Kalman, "Awakening in Chinatown" 1949: 5).

After China and America became adversaries in the Korean War, the dominant organizations in Chinatown further emphasized the community's support for the war against communism. Members of the Cathay Post of the American Legion passed a resolution affirming their loyalty to America and condemning communism at home and abroad. Similarly, the president of the Chinese American Citizens Alliance publicly declared: "If there is war between the Communist countries and the U.S., Chinese-Americans will fight against the enemy of the U.S., regardless of what nationality he is" ("Loyal to America" 1950: 1). The CCBA also wired President Truman and the UN Security Council to protest the possible seating of the Communist delegation.

The *Chinese Press*, meanwhile, stressed the assimilation and the loyalty of Chinese Americans. In February of 1951, it reprinted the words of a white columnist who had recently been in the city. "San Francisco's Chinatown just isn't very Chinese anymore—much less Chinese Communist," he wrote. The author conveyed the signs of acculturation that he observed during his visit: "They sit in soft drink parlors and listen to the Cal-Washington game. . . . The Irish Mission district chose a Chinese-American girl for centennial queen. . . . Chinese families now live all over town, including Nob Hill." The author came to the conclusion that "Instead of being un-American, they are helping us learn the true meaning of Americanism" ("PRESS Inspires Editorial" 1951: 5). The Thanksgiving issue a few months earlier emphasized assimilation with a photograph of a three-year-old Chinese American boy sporting a cowboy outfit. The caption said that the boy was "symbolically turning his back on old Chinatown," and told readers the following:

> As the world is even now facing a choice between opposing ideas and China itself seemingly is lost to the forces of totalitarianism, Chinese-Americans have an unparalleled opportunity to preserve the best of their ancestral values while continuing to be a living testimony to their American way of life. ("Hopalong Cassidy" 1950: 1)

By affirming their anticommunism and their loyalty to America, the community leaders hoped to protect Chinese Americans. The Chinese embassy had already received complaints that Chinese Americans were being "subjected to uncivil remarks and . . . [mistreated] in other ways and that some of their shops . . . [had] been stoned" ("Footnote on Korean War" 1950: 3). Community newspapers noted similar incidents and recommended that Chinese Americans "maintain closer contacts with Caucasian groups and rely on the friendly support of liberal elements and opinion" in order to "offset those with anti-Chinese racial prejudice [from] taking advantage of the situation" ("Chinese Forum" 1950: 4). Some Chinese Americans were afraid that they might be forced to endure a mass internment, especially because the McCarran Internal Security Act authorized the arrest and detainment of suspected spies and saboteurs. To allay such fears, Governor Earl Warren made a surprise visit to the Chinese American Citizens Alliance headquarters to assure everyone that he would give whatever aid he could when called upon. "The situation is grave," he said, "and it may affect the Chinese in America, but Americans of Chinese origin are all American citizens" ("California's Gov. Warren" 1950: 1). Because Warren had been a staunch advocate for Japanese American internment during World War II, his statement directed toward these nonwhite citizens indicated that at least in this case America would not automatically associate Asian ancestry with the Asian enemy abroad.

Chinatown's KMT-dominated political leadership tried to convince the American public that the Chinese people, both in the United States and in China, did not consider PRC government officials to be their legitimate leaders. The president of the CCBA told *San Francisco Examiner* readers that the "Red delegates speak only for Moscow—they do not speak for

China. The fact that China was conquered by force does not mean the people of China have become Communists." In the same article, the head of the Chinatown branch of Bank of America claimed that "the best means by which America can avoid total war lies in the adoption of a program of all-out aid to Chiang Kai-shek, coupled with adequate supervision and an abundance of technical aid ("S.F. Chinese Oppose" 1950: 3). Chinatown leaders thus portrayed the KMT not only as the faction most closely aligned with America's foreign policy objectives but also the real Chinese government in exile. Chinese San Franciscans personally conveyed their anticommunist political views to President Truman as well. A delegation of prominent Chinese San Franciscans met with Truman to pledge the community's support for America's anticommunist war, calling it a "program to restore the liberties to all freedom-loving peoples" ("Chinese Leaders Ask" 1951: 1). Lim P. Lee, then the commander of the Cathay Post and the delegation's spokesman, asked Truman to include Nationalist China as a signatory to the San Francisco Peace Treaty and to give substantial economic and military aid to Chiang's forces.

In January of 1951, anticommunism in Chinatown became formally and publicly institutionalized when the CCBA created what it called an Anti-Communist League. According to Gilbert Woo, this league was affiliated with the anticommunist league in Taiwan, and was therefore transnational rather than homegrown. One of the organization's objectives was to help Americans "differentiate between friend and enemy among the Chinese" (Nee and Nee 1973: 225). To realize this goal, the group furnished the names of suspected leftists and leftist sympathizers to the FBI and the INS. The state agencies took those names and began harassing and in some cases even deporting Chinatown leftists. State repression, therefore, helped silence the dissenting voices coming from the left and further bolstered the dominance of the KMT (Wang n.d.: 415). Reportedly representing all but two or three Chinatown organizations, the Anti-Communist League also asked President Truman to inform the public that the Chinese in America "are not in sympathy of, or affiliated with, the actions or policies of the Communists in China" ("Kung hay fat choy" 1951: 10).

Around the same time that the Anti-Communist League emerged, the Cathay Post of the American Legion and the Chinatown post of the Veterans of Foreign Wars spearheaded another effort to emphasize the loyalty of Chinese Americans. They endorsed the creation of a memorial honoring Chinese Americans who died in World War I and World War II. Within one week, they raised 10,000 dollars to finance the project. When finally completed, the monument stood in St. Mary's Square, directly across from the statue of Sun Yat-sen. Its bronze plaque listed the names of the Chinese American war dead (see Figure 3.3). According to Charles Leong, the memorial, "with its resulting flood of favorable public response and recognition of the Chinese-American sacrifices, stilled the emotional anger against us" (Leong 1976: 53).

For Lim Lee, the gradual nullification of America's anti-Chinese legislation and the Communist victory in China ended the debate over whether Chinese Americans had a better future in China than in America. "Prior to the Peking government," Lee wrote in the *Chinese Press*, "we must admit there were cases of dual loyalty in China and to the United States. . . . Today our future is in this country; our loyalty is to the United States." "Our grandfathers and fathers still had a free China to return to when the going was rough," Lee continued, " . . . but we have no free China to which to retreat. Our future is in this country. . . . This is the Promised Land!" (Lee 1950: 9). Lee's declaration of loyalty to America is undoubtedly genuine, but questions concerning identity and politics within Chinatown were less settled than he claimed. The historical circumstances that created the statue and the war memorial in St. Mary's Square suggest that the linking of the global and the local in Chinatown politics was strong, and the evolution of Chinatown politics in the second half of the century confirm the ongoing relevance of transnational factors. As Michel Laguerre and Jan Lin both point out, even the decline of the KMT in Chinatown politics twenty years later resulted from inseparable global and local factors. Post-1965 immigration and

Figure 3.3 Lim P. Lee holding a drawing of a plaque memorializing Chinese-Americans who lost their lives in both world wars, 1951. San Francisco History Center, San Francisco Public Library.

the normalization of relations between the United States and China increased and emboldened the population that disagreed with KMT political positions. At the same time, the Asian American movement, itself a by-product of antiwar protest and community concerns, led to the rise of social welfare organizations to serve Chinatown (Laguerre 1999: 38–41; Lin 1998: 126–127). Chinese Americans also now worked through American political parties and the different levels of government. By the end of the century, the CCBA was no longer the force it once was. One Chinatown analyst observed:

> They are kind of an anomaly in San Francisco. They are ultraconservative. But, they are not good in local politics. . . . You see, the "Six Companies" [CCBA] is just too closely tied to Taiwan; that is why it cannot move. . . . The majority of Chinatown residents are prounification of Taiwan with China. (Laguerre 1999: 34)

The change in U.S.-China relations may have helped diminish the significance of the CCBA, but other transnational characteristics in Chinatown persist. The history of Chinatown politics in the middle of the twentieth century suggests that ethnic leadership had business and government relationships that were local and transnational in nature. In our time, business between Chinatown and China is no longer restricted to a few brokers, and China's investment in the enclave and Chinatown's business ties to Chinese companies situate Chinatown as an important site of exchange. This process is occurring not only in Chinatowns in the United States but also in Asian ethnic communities throughout the Pacific Rim. Meanwhile, today's ongoing Asian migration, instant global communication, and border crossing in other forms facilitate the continuation of both local and diasporic political identities. Even Chinatown tourism, including walking tours that visit the Sun Yat-sen statue in St. Mary's Square, reinforces how Chinatown is both local and global space.

NOTES

1. William Lawson's speech in Charles Leong papers, carton 1: 40 "Sun Yat Sen." Ethnic Studies Library, University of California, Berkeley, CA.
2. "Chinatown in Transition," unpublished manuscript dated 15 April 1939, p. 14, in Pardee Lowe papers, box 381: "Chinatown in Transition." Hoover Institution Archives, Stanford University, Stanford, CA.
3. YWCA newsletter dated 18 December 1937, p. 9, in Lowe papers, box 382: "Youth Organizations."
4. Down Town Association of San Francisco newsletter dated 22 June 1939 in Lowe papers, box 382: "Chinatown-San Francisco."
5. "Chinatown in Transition," p. 11.
6. "Chinatown Evenly Divided in Attitude toward Civil War," newspaper clipping of *San Francisco Chronicle* article dated 3 May 1949, n.p., in Lowe papers, box 372: "Politics—Chinese."

REFERENCES

Admit the Chinese! (1943, September 9). *People's World*, p. 1.
Art: Statues (1937, November 22). *Time*, 30 (21): 35.
California's Gov. Warren Assures Chinese-Americans (1950, December 8). *Chinese Press*, p. 1.
Chen, Yong (2000). *Chinese San Francisco, 1850–1943: A Trans-Pacific Community*. Stanford, CA: Stanford University Press.
Chiang Stooges Mar Chinatown Double 10 Fete (1949, October 12). *People's World*, p. 2.
China Imports Arrive Here Despite War (1949, May 6). *Chinese Press*, p. 1
Chinese Forum: Chinese in American Today, The (1950, December 15). *Chinese Press*, p. 4.
Chinese Leaders Ask President's Aid in Throwing Off Yoke of Red Tyranny (1951, September 7). *Chinese Press*, p. 1.
Chun, Gloria H. (1993). *Of Orphans and Warriors: The Construction of Chinese American Culture and Identity, 1930s to the 1990s* (doctoral dissertation). University of California, Berkeley, CA.
Chun, Gloria H. (1998). "Go West . . . to China": Chinese American identity in the 1930s. In K. Scott Wong and Sucheng Chan (eds.), *Claiming America: Constructing Chinese American Identities During the Exclusion Era* (pp. 165–190). Philadelphia: Temple University Press.
Chung, Sue Fawn (1998). "Fighting for Their American Rights: A History of the Chinese American Citizens Alliance." Pp. 95–126 in *Claiming America: Constructing Chinese American Identities During the Exclusion Era*, edited by K. Scott Wong and Sucheng Chan. Philadelphia: Temple University Press.
CIO Hits Ban on Chinese (1943, April 16). *People's World*, p. 1.
Decade of Growth for the Chinese-Americans in California, A (1950, November 24). *Chinese Press*, p. 9.
Dobie, Charles C. (1936). *San Francisco's Chinatown*. New York, NY: D. Appleton-Century Company, Inc.
Footnote on Korean War; Local Incidents Reported (1950, December 22). *Chinese Press*, p. 3.
Hoodlums Fail (1949, October 11). *People's World*, p. 2.
Hopalong Cassidy from out of Chinatown, A (1950, November 24). *Chinese Press*, p. 1.
Jesperson, T. Christopher (1996). *American Images of China, 1931–1949*. Stanford, CA: Stanford University Press.
Kalman, Ted (1949, June 24). The Awakening in Chinatown. *People's World*, Section II, p. 5.
Kalman, Ted (1949, October 13). Protection Asked for Meets. *People's World*, p. 1.
Kalman, Ted (1949, October 14). Chiang Spies Here See End of Reign. *People's World*, p. 2.
Kung hay fat choy (1951, February 19). *Fortnight*, 10 (4): 10–12.
Labor Council Backs FR Veto—Beats Move against Racial Bans (1943, June 29). *People's World*, p. 4.
Laguerre, Michel S. (1999). *The Global Ethnopolis: Chinatown, Japantown and Manilatown in American Society*. New York: Palgrave Macmillan.
Lai, Him Mark (1991). "The Kuomintang in Chinese American Communities before World War II." Pp. 170–212 in *Entry Denied: Exclusion and the Chinese Community in America, 1882–1943*, edited by Sucheng Chan. Philadelphia: Temple University Press.
Lee, Lim P. (1940, January). Observations on Elections in Chinatown. *Chinese Digest*, 6 (1), News Supplement, p. 1.

Lee, Lim P. (1950, November 24). A Decade of Growth for the Chinese-Americans in California. *Chinese Press*, p. 9.

Leong, Charles L. (1976). *The Eagle and the Dragon: A Real-Life Chinese-American Story*. San Francisco: Chinese Bilingual Pilot Program, San Francisco Unified School District.

Lin, Jan (1998). *Reconstructing Chinatown: Ethnic Enclaves, Global Change*. Minneapolis: University of Minnesota Press.

Lowe, Pardee (1936, February). Chinatown's Last Stand. *Survey Graphic* 25 (2): 86–90.

Loyal to America. (1950, December 8). *Chinese Press*, pp. 1, 3.

Ma, L. Eve Armentrout (1991). "Chinatown Organizations and the Anti-Chinese Movement, 1882–1914." Pp. 147–169 in *Entry Denied: Exclusion and the Chinese Community in America, 1882–1943*, edited by Sucheng Chan. Philadelphia: Temple University Press.

Nee, Victor G., and Nee, Brett de Bary (1973). *Longtime Californ': A Documentary Study of an American Chinatown*. New York: Pantheon Books.

Peking and the West (1949, October 11). *San Francisco Chronicle*, pp. 1, 9.

PRESS Inspires Editorial on S.F. Chinese-Americans (1951, February 9). *Chinese Press*, p. 5.

Ringer, Benjamin B. (1983). *"We the People" and Others: Duality and America's Treatment of Its Racial Minorities*. New York: Tavistock Publications.

San Francisco (Calif.) Board of Supervisors (1943, August 30). Urging Repeal of the Chinese Exclusion Act. *Journal of Proceedings, Board of Supervisors, City and County of San Francisco*, 38 (38), 1963–1967.

S.F. Chinese Oppose UN recognition of Reds (1950, December 22). *Chinese Press*, p. 3.

Starr, Milton B. (1873). *The* Coming Struggle*: What People on the West Coast Think of the Coolie Invasion*. San Francisco, CA: Bacon and Company.

Wang, Ling-Chi. (n.d.). *Politics of Assimilation and Repression: History of the Chinese in the United States, 1940–1970*. Unpublished manuscript. Ethnic Studies Library, University of California, Berkeley, CA.

Yung, Judy. (1995). *Unbound Feet: A Social History of Chinese Women in San Francisco*. Berkeley, CA: University of California Press.

Yung, Judy. (1999). *Unbound Voices: A Documentary History of Chinese Women in San Francisco*. Berkeley, CA: University of California Press.

Part II

Urban Contraction and Expansion

Economic Restructuring and Governance in De-Globalizing and Globalizing Secondary Cities

Part II

Urban Contradiction and

4 Staggering Job Loss, a Shrinking Revenue Base, and Grinding Decline
Springfield, Massachusetts, in a Globalized Economy

Robert Forrant

> It was there that the sleight of hand lawyers proved that the demands lacked all validity for the simple reason that the company did not have, never had had, and never would have any workers in its service because they were all hired on a temporary and occasional basis. . . . and by a decision of the court it was established and set down in solemn decree that the workers did not exist.
>
> —Gabriel Garcia-Marquez, *One Hundred Years of Solitude*

INTRODUCTION

> It's sad. I didn't realize how much it meant to me, till I think about not going back in there. It's a 36-year habit that's going to be hard to break. I can close my eyes and walk through the building, smelling the cutting oil, hearing the machines.
>
> —Donald Staples, 36-year American Bosch machinist, on hearing the closing announcement on February 4, 1986

On February 4, 1986, at approximately 2:00 p.m., thousands of workers and their families' lives changed in ways they could only begin to imagine. On that day United Technologies Corporation (UTC) ordered the closure of the seventy-six-year-old American Bosch manufacturing plant in Springfield, Massachusetts, capping a nearly thirty-year history of job loss and work relocation from the sprawling factory. The plant was built in 1911 by the Robert Bosch Magneto Company of Stuttgart, Germany, and in the years before it closed workers manufactured precision fuel-injection systems for automobiles, trucks, and tanks. Early on, the four-story plant produced electrical starters and other car parts as well as, for a brief period, radios. Bosch's history represents the quintessence of the story of manufacturing companies in the Connecticut River Valley and across the northern tier of the United States.

For over 150 years Springfield, Massachusetts, stood at the approximate center of a prosperous two-hundred-mile industrial corridor along the Connecticut River between Bridgeport, Connecticut, and Springfield, Vermont, populated with hundreds of machine-tool and metalworking plants and thousands of workers. The valley's richly variegated manufacturing base, deeply rooted in a set of industries that required at their core large numbers of skilled workers, produced economic prosperity well into the twentieth century. Firms relied on these workers to both submit and exchange their ideas to improve manufacturing processes. Continuous skill development was reflected in the widespread recruitment and persistent cultivation of precision machining skills. The arrangement was so successful, according to one account, "that the aggregation of metalworking shops it produced was a model of economic success for 150 years" (Boucher 1994). Through the first half of the twentieth century the valley's machinery builders fostered the growth of a range of industrial districts in Massachusetts such as watches in Waltham, footwear in Haverhill, furniture in Gardner, textiles in Lowell, jewelry making in Attleboro, cutting tools and taps in Greenfield, and metalworking and specialist machine making in Worcester and along the Connecticut River. It would not be hyperbole to call the collection of firms along the river the 'Silicon Valley' of its day, or one of the most advanced manufacturing regions in the world at that time. This early importance of small and secondary New England cities long predated our current attention to them as critical to and in the global economy, politics, and culture today (see Kanna and Chen introduction; Chen and Magdelinskas epilogue, this volume).

Built at the start of the nineteenth century, the Springfield Armory propelled the economy forward. By the early twentieth century, firms like American Bosch advanced the valley's economy further and provided well-paying work for thousands of people. However, starting in the late 1940s, outside investors and manufacturing conglomerates bought many of the area's leading firms, including Bosch. New owners, intent on securing a rapid and high return on investments, were disinclined to nurture the skill base and felt no obligation to the valley's workers or its rich industrial heritage (Best and Forrant 1998).

WHO WORRIES ABOUT BLUE-COLLAR WORKERS?

In the 1980s a dramatic wave of layoffs and plant closings among the Connecticut River Valley's largest machine tool and metalworking manufacturers led to rapid industrial decline and massive dislocation for thousands of the region's best-paid workers. The people who did the work and the trade unions that represented them became extinct. Compounding the situation, there were nearly two decades of hapless efforts by various local and state governments to overcome employment loss and build some sort of sustainable economy. Springfield, Massachusetts, the hardest hit city, fell into serious disrepair. Corrupt officials exacerbated the problem and caused residents, already cynical about their government, to become even more disenchanted with city leaders. Its once-powerful agglomeration of skills and innovative firms depleted, Springfield staggered, nearly bankrupt, into the new century (Forrant and Flynn 1998). Even New England's vaunted education and training system, which once satisfied employers' skills demands, couldn't turn the tide of economic failure. In an earlier era, firms sought out the Connecticut River Valley for its abundance of "high skills and scientific workers." Firms settled in the Connecticut River Valley because

> areas like New England have a large number of firms which change their products and production processes frequently. Change means restructuring, learning new methods, testing, and experimenting. While a company which produces a large volume of output using a well-defined and unchanging production process looks to site its plants in low cost areas with little regard for distance from headquarters, companies which are changing and developing usually must keep a close watch on production. The combination of first, the need for specialized skills and, second, the changing nature of a firm's need for resources helps keep industrial agglomerations together (Hekman 1980).

Springfield and the valley's deindustrialization story exemplifies how globalization and agglomeration gone badly affected hundreds of old industrial cities (see Ryan's chapter in this volume on the decline of the Detroit-Windsor region). When firm managers stopped investing in their skill base and searched for cheaper skills outside the region and the nation, the region's "innovation agglomeration" atrophied.

Recent historical studies of the textile, machinery, metalworking, and plastics industries reveal that a systematic learning process among enterprises and institutions strengthens skills and builds a regional economy's capacity for innovation. Prosperous "learning regions" depended on workers applying their intelligence at work; this facilitated the lifelong learning required for successful knowledge-intensive production (Antonelli and Marchionatti 1998; Capecchi 1997; Herod 1997; Scranton 1997; Wieandt 1994). The proposition held true in greater Springfield for nearly two centuries as new machine tools and knowledge sharing advantaged firms and by extension the city and the region. Skilled workers exchanged their expertise; the idea was to "accept the best and use it to the shop's betterment" (quoted in Malone 1988).

In the early 1950s, President Truman's Council of Economic Advisers warned him that New England firms were turning away from their historical strengths: skill development, technological innovation, and the diffusion of new production methods (Smith 1977). By 1968 even the historically important Springfield Armory succumbed in the wake of defense spending cuts. In response, Springfield's Chamber of Commerce changed the city's nickname from the "Industrial Beehive" to the "City of Homes" (Forrant 1996). The scale of decline, evidenced by the search for the next cheap place to get things made, supplanted the valley's golden age of skill; metal fatigue had set in (Forrant 2009).

In 1982 organized labor demanded that the Massachusetts legislature adopt plant-closing legislation aimed at stopping the job loss. Governor Michael Dukakis established a thirty-eight-member Commission on the Future of Mature Industries made up of leaders from business, labor, government, and academia and charged it with developing "an industrial policy to move the state toward a truly balanced economy . . . in terms of its industrial mix and in its distribution of benefits to the Commonwealth's regions and citizens" (Leff 1986). Three issues overshadowed the commission's work: the types of extra assistance workers caught in closings should receive; the proper economic development role for state government in distressed regions; and whether firms should be required to give their employees and the state prenotification of layoffs and closings. While the commission deliberated, two hundred more firms closed and eighteen thousand additional workers lost their jobs (see Table 4.1). In the end, the commission produced language that only encouraged employers to give workers ninety days' notice of shutdowns. A new Industrial Services

Table 4.1 Layoffs and Closings in Springfield-area Metalworking, 1982–1990

Company	**Status**	**No. of Jobs Eliminated**	**Closure Dates**	**Years in City**	**Peak Emp. since 1960**
American Bosch	Closed	1,500	2/86	80	1800
Chapman Valve	Closed	250	6/86	100+	2700
Columbia Bicycle	Closed	250	6/88	80+	1000
Kidder Stacy	Closed	90	9/89	100+	325
Northeast Wire	Closed	35	1990	22	125
Oxford Precision	Closed	60	9/86	40	120
Package Machinery	Closed	400	9/88	100+	950
Plainville Casting	Closed	65	4/87	65	75
Portage Casting	Closed	60	8/86	36	100
Rafferty Steel	Closed	50	11/85	40	—
Rexnord Roller Chain	Closed	200	6/89	100+	675
Springfield Foundry	Closed	75	4/86	100+	285
Van Norman	Closed	275	10/83	90	1200
Van Valkenberg Plating	Closed	40	7/86	100+	135
Wico Prestolite	Closed	250	3/82	80	675
Atlas Copco	Layoffs	565	1980s	70+	1000
Easco Hand Tool	Layoffs	2,000	1980s	75+	2200
Storms Drop Forge	Layoffs	125	1980s	60+	250

Source: Robert Forrant, *Metal Fatigue* (2009: 166).

Program worked with unions and firms to avert closings, provide retraining if a closing occurred, and help specific communities affected by a large number of closures (Leff 1986).

Bosch and Springfield: Intertwined Histories of Prosperity and Decline

In December 2004, a fire destroyed the former American Bosch plant, which closed in 1986. The morning after the fire, managers at Danaher Tool (formerly Moore Drop Forge and EASCO Hand Tool), one of Bosch's North End neighbors and one of the few remaining metalworking firms in Springfield, announced its closing. Anxious to see the factory's ruins, and anticipating that several ex-Bosch workers would too, I drove to the fire scene. At the destroyed building, it felt like I was attending a friend's wake. Still standing was the once white, now soot-stained front wall of the original factory building with the year 1911, signifying when the factory opened in the city, still visible. Around me stood one hundred former Bosch workers, each one telling stories about the old days.

Memories flooded back as I walked to the far corner of the building and peered through the rubble into the area of the factory where I had once worked. I thought about workmates who showed up early every day, started their coffeepots and argued about the relative merits of local sports teams. People sold donuts and newspapers to raise money for their children's college tuition or some local charity. Small-time gamblers sold betting tickets on athletic events. The fire symbolized the historical disjuncture between generations of working-class prosperity and the grim reality represented by the decaying neighborhood around the plant and Springfield's and nearby Holyoke, Massachusetts' and Hartford, Connecticut's dysfunctional urban centers. In the three cities, decaying neighborhood housing stock, escalating unemployment, rising crime rates, and failing schools were the result of years of neglect, irresponsible governance, and failed national employment and urban revitalization policies (compare to Ryan, this volume, on the community effect of auto plants' closings in Detroit). At numerous intervals over the previous two hundred years, working people flocked to Springfield, Hartford, and the Connecticut River Valley for the well-paying jobs they could find there. Now young people voted with their feet and moved on if they could while working people were trapped in their neighborhoods unable to sell their homes as the regional economy failed them.

Just a few months before the Bosch fire, the Massachusetts legislature and Republican Governor Mitt Romney took the unusual step of creating a Finance Control Board—chaired by the Commonwealth's Department of Revenue Commissioner Alan LeBovidge—"to initiate and implement extraordinary remedies to achieve a long-term solution" ("LeBovidge letter," September 1, 2004) to the city's money mess. (Compare to the political intervention in Detroit in Ryan, this volume.) By the time of the December fire, the control board had tightened its hold over the city's spending decisions. For the control board, the way forward seemed to come at the expense of teachers, school cafeteria workers, firefighters, policemen, and other municipal workers. For the control board, much like the International Monetary Fund in a developing country, Springfield needed to structurally adjust; it needed to shrink its level of public services and focus on spending curbs ("LeBovidge letter," September 1, 2004; Greenberger 2004).

How far have the city and region fallen? In the 1950s and 1960s, Springfield's and most of the region's metalworkers enjoyed relatively good wages, and the city's downtown and ethnically defined neighborhoods benefited. However, the wave of industrial closings provoked the near collapse of many of the city's leading financial institutions. Because of the consolidation of manufacturing and financial services, Springfield's leading capitalists lost ownership and control over capital, and one after another downtown's local banks and department stores closed. The buildings stood empty, monuments to a failed economy. To add to the problem, overly ambitious and ill-conceived riverfront development projects had failed. Elected officials and local and regional economic development organizations and agencies spent years covering over the decline while searching for a new scheme that could breathe life into region. Harvard University Business School guru Michael Porter and creative economy maven Richard Florida offered their high-priced services to the city. However, in 2004 Springfield, the Bay State's third-largest city had a $41M budget deficit; deindustrialization's chickens had come to the empty downtown to roost! Wishful thinking, mismanagement, corruption, a soaring crime rate, and the failure of the region's extensive network of colleges and universities to contribute much time to strategizing over a solution to economic decline had by now turned even the most profound Springfield optimists into doomsayers (Black 2009; Claeson 1996).

Once installed, the control board focused on a handful of symptoms of distress, but it failed to address the historical relationship between the disappearance of well-paying work and the city's plight. No one in a responsible position discussed how the loss of half of greater Springfield's manufacturing plants between 1950 and 1987 accompanied by the loss of 43 percent of Hampden County's industrial employment between 1980 and 2000 affected the city's finances. There can be little doubt that the cumulative impact of the closings and layoffs breached the historical continuity of the valley as a world leader in precision metalworking. And it was equally obvious that new high-wage replacement work had not materialized. Yet, from 2004 to 2009, rather than focusing some attention on job creation and economic development, the control board centered their turnaround strategy on the containment of personnel costs. Failing to focus on growing Springfield out of its budget mess, recovery remained elusive. In April 2007 the executive director of the control board reported that the city "faces a $3.2 million budget deficit for the fiscal year starting July 1 and deficits for the following three fiscal years" if it tries to pay back the original loan by 2009, as the enabling legislation required. The deficit required further cuts in city services and/or a series of new tax- and revenue-raising measures (Ring, *The Republican*, April 5, 2007).

JOBS MATTER

There were at least six good reasons why members of the Finance Control Board, local and state economic development officials, and political leaders needed to make connections between employment loss and Springfield's fiscal situation. First, blue-collar know-how had been a critical source of the competitive advantage of the city and the region and the basis for many business innovations that brought in wealth from outside the region. Second, well-paid industrial workers paid local and state taxes that helped fund schools and fueled growth in the retail, real-estate, and entertainment sectors. Third, workers' secure employment had allowed them to send their daughters and sons to college and helped reproduce the state's vaunted skill base. Fourth, a statewide infrastructure of hundreds of small and medium-sized metalworking, plastics, and precision manufacturing firms received lucrative subcontracts from companies outside the region. Fifth, the wages in services lagged well behind manufacturing wages. And sixth, median household income in western Massachusetts' industrial cities in 2007 dropped well below the state average of $50,502. The numbers: Greenfield, $33,110; Holyoke, $30,441; North Adams, $27,601; Pittsfield, $35,655; Springfield, $30,417 (data from the Commonwealth of Massachusetts Executive Office of Labor and Workforce Development Web site, www.mass.gov).

Springfield's wage figure helps us understand its financial malaise. Deindustrialization resulted in impoverishment or near poverty for thousands of families. In his ethnographic study of Springfield's poverty-stricken neighborhoods, Bjorn Claeson found that between 1960 and 1990 the number of people employed in manufacturing in Springfield declined from approximately 23,000 to 12,000. He concluded:

> The wholesale and retail industry employed roughly the same number of people in Springfield in 1990 as in 1960 (13,563 versus 13,230). The service industry labor market, by contrast, grew during the same period of time; the total number of services in the city increased from 7,659 to 17,524. But, the service and retail jobs that increasingly dominate the labor market pay an average of $7,000 and $17,000 respectively less per year than the average manufacturing job, even accounting for wage reductions in manufacturing. Many people who formerly would have enjoyed stable industrial jobs and comfortable wage lead an existence even more insecure, having been pushed into the city's dangerous informal service economy (Claeson 1996: 49).

Whereas manufacturing employment fell across Massachusetts and in Hampden County between 2000 and 2004, there was some growth in lower-wage services employment as well as in some engineering and business-services categories. Massachusetts lost metalworking employment: jobs fell in this category from 197,000 jobs in 2000 to 167,656 in 2004, a 15 percent falloff. Hampden County lost 16 percent of its employment over the same period and at the start of 2006 there were slightly fewer than ten thousand jobs in this category, a far cry from the twenty-five thousands jobs present as late as the early 1980s. Across the Commonwealth these jobs were replaced by low-wage employment such as in restaurants or employment in some categories that paid reasonably well but required a far different education than the one most unemployed metalworkers had. Figure 4.1 reveals what happened to Massachusetts's manufacturing and services employment from 1997 to 2004.

As the manufacturing employment crisis worsened, Springfield's leaders wrongly assumed that Massachusetts' boom in electronics, finance, and biotechnology would right the ship. But by late 2001, Hampden County's employment stagnated. Even though the Bay State added almost five hundred thousand service-sector jobs over the 1980s and 1990s, the majority of the well-paying ones were inside the Route 495 beltway, much closer to Boston than to Springfield. According to the *Boston Globe*'s Charles Stein, "The new economy never made it this far outside the Massachusetts Turnpike." Stein summarized, "A lot of middle-class people left for better economic opportunities, while the number of poor people grew steadily over the past two decades. This shift helped make Springfield one of the poorest cities in Massachusetts" (Stein 2004b). Springfield's reported unemployment rate of 8.5 percent was a lot higher in several of its Hispanic and African American

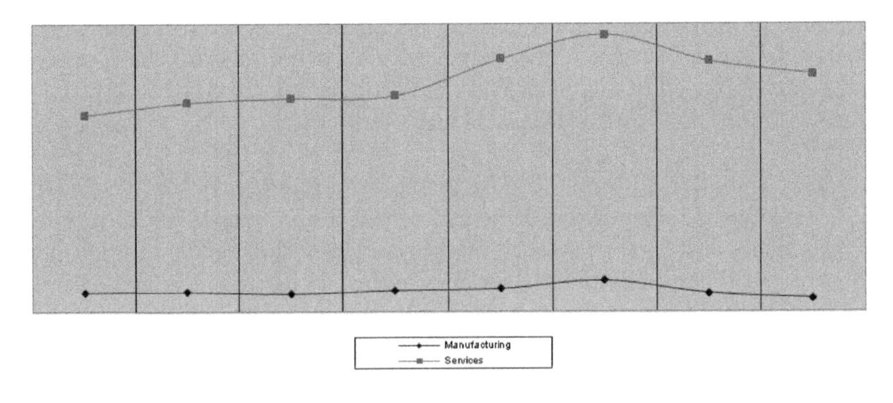

Figure 4.1 Massachusetts manufacturing and services employment, 1997–2004.
Source: Massachusetts Executive Office of Labor and Workforce Development, Current Employment Statistics, http://lmi2.detma.org/lmi/lmi_ces_a.asp
(Accessed between 2007 and 2010).

neighborhoods. The public-health infrastructure got smaller while the city's quality of life deteriorated dramatically as thousands of streetlights were dimmed, fire stations closed, and police officers and teachers were laid off.

To reiterate, the vaunted service economy produced too few well-paying, full-time jobs. In Massachusetts, the gap between average manufacturing and services wages did not close appreciably. For the country, "Industries ranked in the bottom fifth for wages and salaries have added 477,000 jobs since January 2004, while industries in the top fifth for wages had no increases at all. . . ." (Andrews, August 9, 2004: C1; Stein, 2004a: 1). In 2003, three of the largest occupations adding jobs in Massachusetts—cashiers, food preparation and serving workers, and waiters and waitresses—paid average wages below $10 an hour. In the same year, nine of the fifteen largest occupations in the Bay State reported average wages of less than $15 an hour. "Given the high cost of living in Massachusetts this proliferation of low-wage jobs is a major public policy challenge. . . ." (Brenner 2007: 24). According to figures released by the commonwealth, in Hampden County, manufacturing employment fell from 34,301 jobs in 1998 to 28,000 in 2004, while service jobs increased from 86,500 to nearly 96,000 (Massachusetts Executive Office of Labor and Workforce Development, www.mass.gov). Over the period, service-sector wages barely registered an increase over $650 a week, while the remaining manufacturing jobs averaged $850.

Commenting on August 2004 national employment data, *New York Times* business reporter Louis Uchitelle noted that layoffs are "more frequent now in good times and bad, than they were in similar cycles a decade ago" (Uchitelle, *New York Times*, August 2, 2004: C2). Across the country, the almost 60 percent of laid-off workers who found a new job earned less money, compared with about 50 percent of workers who went through the same experience in the early 1990s. In other words, both the number of new jobs and the wages of jobs continued to moderate, so that even when employment growth occurred, the wealth base eroded. This is what transpired in Springfield. Absent tax increases or some infusion of new revenues through a state or federal stimulus program, public services will be further curtailed. In addition, employment dependent on disposable income, like restaurants, will shrink. For a nation as rich as the United States, one-quarter of all jobs paid at or below poverty wages; for blacks, the figure was 30.4 percent and for Hispanic workers 39.8 percent. For black women the figure was 33.9 percent and for Hispanic women, 45.8 percent (Uchitelle, *New York Times*, August 2, 2004: C2).

"A COMATOSE PATIENT ON LIFE SUPPORT": THE FINANCE CONTROL BOARD, 2004–2009

We return to Springfield, Massachusetts. Several Springfield business associations, including Future Springfield, the Springfield Taxpayers Association, and Springfield Central, Inc., unsuccessfully attempted to lead a business revival in the 1980s and 1990s. So-called public-private partnerships were going to save the day but they failed to deal effectively with the city and the region's structural difficulties. Two of the city's leading retailers—Forbes and Wallace and Steigers—vacated their multistory downtown department stores in 1976 and 1994, respectively, no match for the large indoor malls that sprang up along Interstate Route 91. For years, Springfield's recovery rested on a $110 million riverfront development project, anchored by the new Naismith Memorial Basketball Hall of Fame, and the renovation of its downtown civic center. Entertainment complexes, riverboat gambling, and interactive museums were projected for the riverfront to supplement activities at the hall. In other words, tourism dollars were to spark a revival. However, the Basketball Hall of Fame failed to generate the visits that consultants anticipated nor is there any evidence in 2011 that well-paying jobs have materialized along the riverfront, which is cut off from the rest of the city's commercial downtown by a six-lane raised highway and railroad tracks. In 2000, nearly one-quarter of downtown office buildings were vacant. In the early 2000s, Springfield's downturn persisted despite a noticeable economic expansion elsewhere in Massachusetts. According to Robert Nakosteen, a professor of economics and statistics at the Isenberg School of Management at the University of Massachusetts Amherst, "All the wealth just moved out of the city (Goldberg 2005, 1; Nakosteen quoted in Gorlick, *Associated Press*, June 4, 2004).

Yet, the city's employment and revenue problems were not taken seriously, even in the mid-1990s, when it received a $21 million state loan to meet payroll and sold its municipally owned hospital to a private company to cover additional budget shortfalls. In 2004, with the city's debt impossible to ignore, and bankruptcy looming, the Finance Control Board entered the picture, focused on bringing costs in line with projected city revenues by cutting essential services and reducing personnel costs. In September 2004, control board chairman LeBovidge summarized the situation this way: "It is clear that an integral part of the recovery plan for the city of Springfield must include work rule changes, benefits restructuring and take-home pay reductions for municipal workers" ("LeBovidge letter," September 1, 2004). As the crisis worsened, *Republican* reporter Dan Ring wrote, "Although Springfield is the third-largest city in the Commonwealth out of 351 municipalities it has the lowest bond rating, the fourth-lowest income per capita, the second-lowest property values and the highest non-residential property tax rate in the state" (Ring, *The Republican*, June 16, 2004: 1). Things had gotten so bad that "even the trees were falling—their dead limbs crashing onto parked cars and into houses, spurring lawsuits against the city that could not afford to cut the trees down" (Schweitzer, *Boston Globe*, July 1, 2009: B1).

City officials negotiated a relief package with the Romney administration and the legislature. The governor offered approximately $51 million in interest-free loans to cover the shortfall in the $437 million FY 2004 budget, so long as the city agreed to let a state-appointed control board make future spending decisions. Attached to the proposal was the suspension of collective bargaining for the city's unions. Ken Donnelly, secretary treasurer of the Professional Firefighters of Massachusetts, accused the governor of "trying to break the unions. I haven't seen anything this bad in 32 years." Timothy Collins, president of the Springfield Education Association, said that the control board plan represented "the lowest day of my life when you have a mayor and a governor stripping us of our collective bargaining rights. It's almost un-American. Shame on the governor and shame on the mayor." Over the first four years of the control board's existence the Springfield school district "lost 1,325 experienced teachers to better-paying jobs in nearby communities" and "teachers, along with police, firefighters and members of other municipal unions forfeited much of the cash due them from a wage freeze and dropped lawsuits as a condition of settling their contract disputes. . . ." (Ken Donnelly and Timothy Collins, quoted in Tyman, *Associated Press*, July 16, 2004; DeMarco and Richardson, *Boston Globe*, June 6, 2004: B3; Plaisance, *The Republican*, June 27, 2009: 1).

The city now received a $52 million interest-free loan to be paid back by 2012. The $52 million figure matched what the city was owed in delinquent property taxes. With the loan came the Finance Control Board (FCB) to take over fiscal management of the city. The board's term was extended to 2009 in mid-2007 when it became apparent that Springfield wouldn't be able to pay the loan back by 2012, a requirement of the enabling legislation. And despite the arguments from union leaders that job cuts and wage freezes would negatively impact the provision of essential protective and educational services and cause residents to leave the city, the board instituted a wage freeze. Teachers' previously negotiated pay increases were held back, causing the exodus of nearly 250 teachers during the 2004–2005 academic year. In early 2005, Springfield was referred to as "a city under siege" (Goonan, *The Republican*, September 3, 2004: 1; Gorlick, *Associated Press*, June 4, 2004; Tantraphol, *Hartford Courant*, April 17, 2005: 1).

In one of its first public statements, the Board informed residents that

> No solution to the city's fiscal crisis can be achieved without a substantial reduction of personnel costs and expenses. It is clear that an integral part of the recovery plan for the city of Springfield must include work rule changes, benefits restructuring and take-home pay reductions for municipal employees (quoted in Goonan, *The Republican*, September 3, 2004: 1).

Eric Kriss, one of Governor Mitt Romney's chief financial advisors, described the board as "a tool to help the city recover financially" and contended that a turnaround could occur only with the goodwill and effort of municipal employees (Kriss, quoted in Goonan, *The Republican*, September 3, 2004: 1). But with 20 percent of the city's workforce cut between 2002 and 2004, and the private-sector job base continuing to shrink, it was difficult to imagine why city workers would go along with the FCB's plans. Outside observers found it difficult then to see where any badly needed economic stimulus would come from; in 2009 the jury remains out. In September 2004, *The Republican* reminded its readers: "While we agree that work rule and benefit changes need to be made regarding city employees, we vigorously oppose the reduction of their wages and think it would be unconscionable to do so." For good measure, the newspaper's editorial added, "The Control Board should be working for a surgical plan to restore the city's finances, not a hatchet job that leaves the city as nothing more than a comatose patient on life support" (*The Republican*, September 3, 2004: 16).

WHERE TO AFTER FIVE YEARS OF FINANCE CONTROL BOARD (MIS)GOVERNANCE

For most of the last two centuries the Connecticut River Valley, with Springfield as its leading city, related to the rest of the country and the world as a stellar manufacturing center, its metalworkers and machinery builders fueling the nation's industrial revolution. In historical succession, industries including textiles, paper, shoes, rifles and handguns, industrial machinery, aircraft engines, and computers generated spectacular wealth and advanced workers' living standards. Metalworking growth stemmed from three related factors: continual innovations in product design and development stimulated by the Springfield Armory; a nucleus of locally owned, collaborative machine-tool builders and precision metalworking firms whose expertise provided the region with the first-mover benefits of any technological breakthroughs; and the base of skilled workers performing the precision machining required to turn out world-class products.

The original catalyst for economic development in the Connecticut River Valley was the Springfield Armory. Its central objective was the development of precision engineering and its use in gun production. Its willingness to diffuse technical knowledge to its contractors spread best practices, and its ability to attract and train skilled mechanics laid the foundation for the continual reproduction of a highly skilled workforce. In turn, this served as a repository of knowledge, the means of incorporating new technical information, and the source of highly innovative new business start-ups. Burgeoning expertise in machine-tool technology provided the basis for a capital-goods sector, which, together with emerging industries, created a diverse manufacturing base. Over the last half of the nineteenth century and the first sixty-five years of the twentieth century, firms and workers up and down the Connecticut River Valley benefited from learning processes within networks of machine builders, small specialist engineering shops, education and training institutions and final goods producers. Prosperity was assured.

The productive system reached its zenith during the Second World War. Thereafter, Springfield and the region were caught up in the accelerated pace of globalization, and the skill base was no longer a sufficient magnet to preserve and increase well-paying work. Once locally owned firms changed hands, their assets were globalized, and the region's ability to shape and reshape its economic future slipped away. Springfield scrambled to save what jobs they could, offering corporations financial inducement to stay or move in. Springfield and nearby Hartford, Connecticut, even engaged in a ludicrous 'border war,' each city's mayor offering inducements to firms to move employment approximately thirty miles north or south along Route 91. While politicians fiddled, the skill base cultivated up and down the Connecticut River Valley for over a century disappeared. East Hartford, Connecticut, and Springfield—once home to major industries—suffered years of falling living standards and sharp population losses as a result of the collective failure to develop a new 'Armory,' a new catalyst for sustainable prosperity.

For much of the twentieth century a strong shop-floor skill base, combined with innovative and forward-looking employers, provided the region with a competitive advantage for close to 150 years. Symptoms of decline were evident by the early 1950s as the rate of adoption of new technology and the rate of new product development slowed, as skill became a lower priority for firms, as work was relocated to the South and overseas, and as firms turned to short-run financial gain rather than long-run modernization. Increasing competitive failure and the growing interregional and international relocation of production to low-wage areas resulted in progressive plant closures and mass layoffs which undermined the Valley's position as a world leader in precision metalworking. None of the measures taken arrested the horrible slump. Not the intense belt-tightening, or the long-term freeze on teachers' wages, or the too-high-priced academic gurus, still less the fanciful notions that tourism would grow a plentiful supply of 'armory-like' well-paying jobs.

Cooperation among the region's trade association, trade unions, educational institutions, and supportive state and federal agencies was essential for meaningful job creation to occur. But as work disappeared, the opportune moment for cooperation slipped away and thereafter only infrequent lip service was paid to organizing a valley-wide campaign against runaway shops and lost jobs. This left what Herod refers to as the "artificial spatial divides between the workplace and the broader community" intact and the top-down nature of the control board hampered any community-wide conversations about how to regrow the employment base (Herod 2001: 268). Springfield's Finance Control Board stripped residents' voice from helping to rebuild their city. Years into the control board's rule, unions representing teachers, police, and firefighters still fought through the courts for their legally negotiated pay increases. A twenty-three-year-old forklift operator expressed the feelings of many residents: "Just get it over with. To know your city is going broke—it's time to move out of here. There's no opportunity here" (quoted in Ebbert, *Boston Globe*, January 17, 2006: B1). Tim Collins, president of the Springfield Education Association, and an outspoken critic of the control board, summed up events thusly:

> These political leaders have starved this city into this situation, so they could put forward their Draconian agenda. And the leadership of the Legislature is letting the city fail because they want to give Romney a black eye as he runs for president (quoted in Ebbert, *Boston Globe*, January 17, 2006: B1).

In July 2007, Democrat Deval Patrick was elected governor and put several new members on the board, and for the first time the board talked about how it might stimulate economic development. Calls were made for the colleges and universities to lend their expertise to the monumental task of finding several new "engines of prosperity" for the valley, ones that might match the skills and innovation that had previously propelled the regional economy forward. But as of July 2009, little has been accomplished on this front.

To conclude, an interesting question is from which direction did the challenge of globalization come? The much larger history of the Connecticut River Valley productive system shows that in its generative phase its machine-tool industry posed a global challenge rooted in its production and innovative capabilities, which came directly from its success in world markets and indirectly from its transformative effect on American manufacturing. In its degenerative phase the productive system faced a global challenge mainly from Japanese competition based on production and innovation principles that had once been its hallmarks. It also faced a challenge from the globalizing activities of U.S. corporations that used world markets to outsource production, and as the demands of shareholders mobilized on a global scale, focused their managers' attention on short-term gain based on downsizing, deskilling, plant closures, cutbacks in research and development, and the neglect of investment in physical and human capital.

In the end, the challenges and impacts of globalization on older industrial cities depend on the rules of game. The issue here is that there is a direct conflict between the logic of the market as idealized by corporate liberalism and incorporated in Anglo-American capitalism and the logic of production as revealed by historical studies of industrial districts like the one in the Connecticut River Valley. The former has concentrated power in corporate hands, guided by stock market pressure to prioritize the short-term distribution interests of shareholders, whilst the latter requires the diffusion of responsibility to all stakeholders and a concentration on the longer term to allow this empowerment to bear fruit. Whether the challenge of globalization is constructive or destructive will depend on which of these routes is followed, although the present directional signs in Springfield, Massachusetts, are not encouraging.

REFERENCES

Andrews, Edmund (2004). "It's Not Just the Jobs Lost, but the Pay in the New Ones." *New York Times*, August 9: C1.

Antonelli, Cristiano, and Roberto Marchionatti (1998). "Technological and Organisational Change in a Process of Industrial Rejuvenation: The Case of the Italian Cotton Textile Industry." *Cambridge Journal of Economics* 22: 1–18.

Best, Michael, and Robert Forrant (1998). "Community-Based Careers and Economic Virtue: Arming, Disarming, and Rearming the Springfield, Western Massachusetts Metalworking Region." Pp. 314–330 in *The Boundaryless Career: A New Employment Principle for a New Organizational Era*, edited by Michael B. Arthur and Denis M. Rousseau. Cambridge, UK: Oxford University Press.

Black, Timothy (2009). *When Heart Turns Rock Solid: The Lives of Three Puerto Rican Brothers On and Off the Streets*. New York: Pantheon Books. Boucher, Norman (1994). "A Natural History of the Connecticut Valley Metal Trade." *Regional Review* (Winter): 6–12.

Brenner, Mark (2007). "The Economy: A Growing Divide with Uneven Prospects." Pp. 11–32 in *The Future of Work in Massachusetts*, edited by Tom Juravich. Amherst: University of Massachusetts Press.

Capecchi, Vittorio (1997). "In Search of Flexibility: The Bologna Metalworking Industry, 1900–1992." Pp. 381–418 in *World of Possibilities: Flexibility and Mass Production in Western Industrialization*, edited by Charles Sabel and Jonathan Zeitlin. New York: Cambridge University Press.

Claeson, Bjorn (1996). *The System Feeds on Us: An Ethnography of Poor People and Elites in a New England City*. PhD dissertation. Baltimore: Johns Hopkins University.

DeMarco, Peter, and Tyrone Richardson (2004). "Bumpy Road Is Seen for Springfield." *Boston Globe*, 6 June: B3.

Ebbert, Stephanie (2006). "Springfield Edges toward Fiscal Abyss." *Boston Globe*, January 17: 18.

Forrant, Robert (2009). *Metal Fatigue: American Bosch and the Demise of Metalworking in the Connecticut River Valley*. Amityville, NY: Baywood Publishing.

Forrant, Robert (1996). "Skilled Workers and Union Organization in Springfield: The American Bosch Story." *Historical Journal of Massachusetts* 24: 47–67.

Forrant, Robert, and Erin Flynn (1998). "Seizing Agglomeration's Potential: The Greater Springfield Massachusetts Metalworking District in Transition, 1986–1996." *Regional Studies* 32: 209–222.

Goldberg, Marla (2005). "Attendance Dips, Hope Rises at Hall." *Sunday Republican*, April 17, 2005: 1.

Goonan, Peter (2004). "Current Tax Collection Bodes Well." *The Republican*, September 3: 1.

Gorlick, Adam (2004). "Springfield Bailout Bill Stalls Due to Union Concerns." *Associated Press*, June 4.

Greenberger, Scott (2004). "State Eying Municipal Bailout: Springfield Facing $20M Budget Deficit." *Boston Globe*, May 29: B1.

Hekman, John S. (1980). "The Future of High Technology Industry in New England: A Case Study of Computers." *New England Economic Review*, January/February 1980: 5–17.

Herod, Andrew (1997). "From a Geography of Labor to a Labor Geography: Labor's Spatial Fix and the Geography of Capitalism." *Antipode* 29: 1–31.
Herod, Andrew (2001). *Labor Geographies: Workers and the Landscapes of Capitalism*. New York: The Guilford Press.
LeBovidge, Alan (2004). Letter from Alan LeBovidge, chairperson of the Springfield Finance Control Board, to Eric Criss, Secretary, Massachusetts Executive Office of Administration and Finance, September 1.
Leff, Judith (1986). *The Plant Closing Debate in Massachusetts.* Harvard Business School Case Study 9–386–173: 1–23.
Malone, Patrick (1988). "Little Kinks and Devices at the Springfield Armory, 1892–1918." *Journal of the Society for Industrial Archeology* 14 (1): 59–76.
Plaisance, Mike (2009). "5 Years after Finance Control Board Came to Rescue Springfield, City Has $34.5 Million in Reserves." *The Republican*, June 27: 1.
Republican, The (2004). "Finance Control Board Too Quick on the Draw," September 3: 16.
Ring, Dan (2004). "Finneran Downplays Bankruptcy." *The Republican*, June 16: 1.
Ring, Dan (2007). "Springfield Needs More Time to Repay Loans." *The Republican*, April 5: 1.
Schweitzer, Sarah (2009). "Springfield's Overseers Leave a City in the Black." *Boston Globe*, July 1: B1.
Scranton, Philip (1997). *Endless Novelty: Specialty Production and American Industrialization, 1865–1925.* Princeton, NJ: Princeton University Press.
Smith, Merritt Roe (1977). *Harpers Ferry Armory and the New Technology: The Challenge of Change*. Ithaca, NY: Cornell University Press.
Stein, Charles (2004a). "Wages Don't Figure in Rebound." *Boston Globe*, May 5: 1.
Stein, Charles (2004b). "Almost a Ward of the State." *Boston Globe,* June 18: C1.
Tantraphol, Roselyn (2005). "Springfield: A City Under Siege." *Hartford Courant*, April 17: 1.
Tynan, Trudy (2004). "With State Aid Promised, Springfield Now Faces the Hard Part." *The Associated Press*, July 16.
Uchitelle, Louis (2004). "Layoff Rates at 8.7%, Highest Since 80s." *New York Times*, August 2: C2.
Wieandt, Axel (1994). "Innovation and the Creation, Development and Destruction of Markets in the World Machine Tool Industry." *Small Business Economics* 6: 421–437.

5 From Cars to Casinos

Global Pasts and Local Futures in the Detroit-Windsor Transnational Metropolitan Area

Brent D. Ryan

INTRODUCTION

Over the past twenty years the topics of globalization and the global city have come to dominate the intellectual dialogue on international planning and development. Related issues like poverty, neocolonialism, sustainability, and even 'development' itself have all been subsumed into a term whose totalizing implications are matched by the breadth of its utilization in recent research, the best-known of which are so highly visible and widely cited as to hardly require specific identification.

The debate over global city status, or 'globality,' with its implications not only of the inevitability and implied desirability of global city membership, or at least of participation in the growing network of global cities, has neglected numerous aspects of urbanization in the transnational context. The implications of this neglect are as meaningful as the aspects themselves have been unheralded. Chief among these are very real concerns with the desirability, or meaning, of globality; whether or not the attainment of global city status can or should be seen as equivalent to a city's attainment of normative goods like democracy or sustainability. There is substantial and growing cautionary evidence to show that urban globality does not guarantee urban virtue, at least as it pertains to universally considered values like democracy and human rights. The absence of fundamental human values from many global cities could be interpreted as calling into question the very meaning of globality, or at least of demanding that the definition be broadened to encompass human as well as economic values.

Reinforcing the presumed desirability of globality is its seemingly teleological aspect. The progression toward membership in the club of global cities is viewed as an inevitable, irreversible progression in a manner not dissimilar to that espoused by the advocates of modernity-driven 'development' from the 1960s through the 1980s (Rist 1997). The inevitability of the progression to global city status is not only ironic considering the particularity and privilege accorded to owners of that status, which of course mitigates against any substantial broadening of membership, but is deeply married to the seeming inevitability of the progression of globalization itself—an economic, social, and physical interconnectedness that is sweeping the world.

This chapter is written to critique this dominant perspective by illustrating the means by which a particular transnational metropolitan area, that of Detroit-Windsor (USA-Canada), is operating in a manner precisely opposite to the dominant teleological trajectory projected by the advocates of globalization. Detroit-Windsor is a swiftly *de-globalizing* region—a binational metropolitan area whose global dominance has been shrinking for the past several decades. Once a leader in global automobile manufacturing and the home, as of 1955, of three of the largest five corporations in the world (*Fortune* 2008), Detroit-Windsor has in the succeeding five decades undergone a shocking deindustrialization that has devastated the economy and landscape of much of the region (compare to Forrant's discussion on the severe decline of the Connecticut River Valley in this volume). Even more profound is the region's transformation from a global leader, if not a global city,[1] to what can only be described as an increasingly marginal role on the world economic stage. In spite of its location straddling two of the world's largest and wealthiest economies (United States and Canada), Detroit-Windsor stands as a signal example of de-globalization, or localization, in a world that seems to be obsessed with the opposite transformation. The global past of Detroit-Windsor is over; what lies ahead will almost certainly be a local future.

Whereas the deindustrialization of Detroit-Windsor has been well documented (Sugrue 1996; Thomas 1997), the planning-level responses to it, carried out substantially at the municipal scale, have rarely been placed into the macroeconomic context that they have ultimately sought to correct. Urban, regional, and state officials in both Michigan and Ontario, faced with large-scale economic shifts whose origin and remedy lie far beyond local control, have sought, through a combination of subsidy and incentives to private developers, to revitalize the economic, social, and physical landscape of the region through the construction of large-scale urban developments. The purpose of these megaprojects is as much to revise and rehabilitate the deteriorating image and physical fabric of the city-region as it is to regenerate an economy whose industrial basis is unrecoverable in a postindustrial, service-based transnational society. (See Forrant's discussion on the political response in the Springfield area, this volume.)

The megaprojects constructed on both sides of the border have a common basis in consumption rather than production. In each development, gambling, entertainment, and leisure are variously presented as 'answers' to an economic crisis whose origins lie in declining wealth, not in its creation. The employment created by the megaprojects is marketed as a substitute for that formerly provided by industrialization, yet the economic foundation of this new economy is dependent on capturing consumption and capital from the very regional residents who have endured the difficult decline of recent decades. For all intents and purposes, the new megaproject economy is a parasitic one, feeding on the embers of the industrial economy by capturing fragments of the accumulated capital generated over decades through industrial production and in possession of the region's residents. The megaproject strategy attempts to capture this capital through local consumption rather than permitting it to be reinvested or dissipated in other regions of the United States/Canada or even farther afield. Yet the majority of the profits generated by this new parasitic growth are simply reaccumulated outside the region by the mammoth corporations and wealthy individuals in control of the megadevelopments, many of whom are located in Nevada.

A DECLINING INDUSTRY IN A DECLINING REGION

In recent decades the relative decline in the dominance of the "Big Three" American-owned automobile corporations (General Motors, Ford, and Chrysler) has been both progressive and relentless. Hampered by competition from agile, efficient competitors in Japan and by their own patent inability to respond with innovations of their own, the Big Three have seen their dominance of the U.S. auto market shrink from a historic high of 91 percent in 1965 to 44 percent in 2009 (Ward's Automotive Group 2010). Since 2000 their decline has been even swifter, with the Big Three's 2010 market share dropping 20 percentage points in ten years to 45 percent (Ward's Automotive Group 2010) in both Canada (Guilford 2010) and the United States (Keenan 2010) and their worldwide market share dropping from 36 percent in 2000 (OCIA May 2001) to only 26 percent in 2010 (Petry 2010, Zacks Equity Research 2010).

This dramatic decline has of course had devastating consequences for automobile-sector employment, much of which has historically been centered in Michigan in the United States and in western Ontario in Canada. In the United States alone, Big Three employment shrank from 435,000 in 2000 (Center for Automotive Research 2008a) to only 171,200 in 2010 (see Table 5.1 for more detailed data and sources). At the same time, the dramatic rise in automobile ownership and production of Big Three automobiles outside of the United States has led to the Big Three's employment abroad rising from 249,622 in 2000 to 276,800 in 2010 (see Table 5.1).

Despite the transfer of much of their production abroad in a desperate attempt to maintain profitability through much lower labor costs, the Big Three's losses in market share and steadily increasing costs in North America (100% increases for materials and much higher than that for employee benefits since 2000) have led to steadily increasing losses for the companies. None of the Big Three were profitable during the period spanning 2000 to 2008 (Center for Automotive Research 2008a); their cumulative losses in 2008 totaled over $104 billion (Orbis Company reports 2010a, 2010b, 2010c, 2010d, 2010e).

Table 5.1 The Big Three's Employment in the United States and Overseas, 2000 and 2010

				Detroit 3 Total
	GM	Ford	Chrysler	
Nationwide				
2000		165,081°°		**425,132**
2010	*	61,200**†	*	**171,200†**
North America				
2000				
2010		73,000††		
Worldwide				
2000	195,374°	352,380°°	127,000°°°	**674,754**
2010	196,000±	198,000±±	54,000†††	**448,000**
Abroad				
2000				**249,622**
2010				**276,800**

Source: These figures are from †Naughton (2010); °Shepardson (2009); ±Guilford (2010); °°Ford Motor Company (2002); ††Targeted News Service (2010); ±±Krisher (2010); °°°McCracken (2005); †††Hoovers (2010)

The 2008 economic crisis badly damaged the already declining Big Three. Both Chrysler and General Motors underwent Chapter 11 bankruptcy proceedings in 2009. Chrysler emerged from bankruptcy only after the sale of most of its assets to an Italian-owned consortium led by the Fiat Group (Naughton, Green, and Welch 2010). This emergence also required $6.6 billion from the U.S. federal government to permit Chrysler to operate as a "debtor-in-possession" of the company's facilities, permitting it to continue operating even when bankrupt (*New York Times* 2010b). General Motors required similar assistance, receiving $9 billion in aid from the federal government even as it could only emerge from bankruptcy by selling its assets to a new company 61 percent owned by the federal government. The new "General Motors Company" promptly began the process of closing thousands of surplus dealerships and dispensing with peripheral brands like Saturn, Pontiac, and Hummer (*New York Times* 2010a). Following dramatic plant shutdowns from all three automakers, profits began to return for Ford in 2009 (Hoover 2010) and 2010 (Tweh 2010), GM (*TendersInfo 2010*), and Chrysler (*Associated Press* Financial Wire 2010).

It does not take sophisticated economic analysis to see that the losses sustained by the Big Three during most of the past decade cannot be maintained; no company can lose money forever, and North America and the Detroit-Windsor region need to face the very real possibility that one or more of the Big Three will cease to exist within the next several years, barring survival through ownership by or even merger with another company. This desperate strategy is one that Chrysler already pursued from 1998 to

2007 when it was owned by the German manufacturer Daimler AG. (Since 2007 Chrysler has again existed as an independent company.)

With much automobile production and employment historically centered in Michigan and western Ontario, the worldwide decline of the Big Three has been even more traumatic at the local level. In Michigan alone, automobile-related employment has dropped nearly 50 percent in eight years from just over 320,000 in 2000 to 125,600 in April 2010. This figure includes related supplier industries outside of the Big Three themselves (U.S. Department of Labor Statistics 2010a, 2010b). The Windsor, Ontario, area, with a much smaller employment base to begin with, has lost almost 18,000 jobs during approximately the same period (De Bono 2007, Gomes 2008). Ironically, the transfer of Canadian automobile production to elsewhere in Ontario, versus the transfer of U.S. automobile production outside of Michigan to other states, led to Ontario surpassing Michigan in vehicle production as of 2004, when both the state and the province produced 2.6M cars. Yet this figure in turn represents a relative decline from 1999, when Ontario produced 2.9M cars and Michigan 3.1M. Accompanying this decline in production has been a flurry of plant closings. In Michigan alone, eighteen plants have been closed or announced for closure from 2005 to 2011. Ontario, on the other hand, has lost only two plants (Center for Automotive Research 2008b).

The Detroit-Windsor metropolitan area remains heavily dependent on automobile manufacturing, but the industry's precipitous decline has caused significant economic distress (see the economic distress in Springfield; Forrant, this volume). With 10 percent of its workforce directly dependent on the Big Three, Windsor was in 2009 the Canadian city with the highest level of unemployment (12.4 percent) in the country (Statistics Canada 2010). The Detroit metropolitan area, similarly dependent on automobile-related employment, suffered as of April 2010 from the highest unemployment rate for large municipalities in the United States: 14.8 percent versus a national level of 9.9 percent (U.S. Department of Labor Statistics 2010b). During the first half of 2009, the State of Michigan Labor Department estimated that the state was losing 32,000 jobs per month, up to 18,000 of which were due to the auto-industry restructuring process alone (Buckwalter Berlooz 2010).

THE SHUTTERED PLANTS

In the face of such a tremendous shrinkage of the automobile economy, any traditional economic development measures that might be considered—particularly the recruitment, retention, or reopening of automobile-related manufacturing facilities—attains a sort of poignant futility. Many of the closed plants are sure never to reopen: like the older facilities of the early and mid-twentieth century located within the city limits of

Detroit and abandoned from the 1950s to the 1970s, they will eventually become the white elephants of their locality, with demolition and reuse by even small amounts of office activity an increasingly attractive option to the continued deterioration of the enormous, abandoned facility (Ryan and Campo 2007).

Some limited reopenings have occurred through subsidies, but they have been outnumbered by closures. In 2008, Ford returned some 300 jobs to its shuttered Essex Engine Plant outside of Windsor. This is fewer jobs than were lost (900) when the facility closed in 2005, but in the face of the declining regional economy the facility's smaller-scale continuance, achievable only with a public subsidy of $80 million, was hailed by Canadian politicians as an economic development success (Vander Doelen 2008a). Yet in 2009, Ford closed another section of the plant, the Nemak Essex Aluminum Plant, causing the loss of another 400 jobs. This part of the plant, which manufactured engine blocks, cylinder heads and other Ford parts, was made redundant by a parallel facility in Mexico, or as Ford put it, by "industry cost-reduction pressures and global capacity optimization" (Schmidt 2009). Plans for this area of the plant remain uncertain.

Elsewhere in Windsor, the news has been equally grim. Another major employer, the General Motors Windsor Transmission Plant, was announced for closure in May 2008, due both to the simple obsolescence of the plant's principal product—four-speed automatic transmissions—and to the obsolescence of the plant itself. The plant will close in the middle of this year (2010), putting 1,400 employees out of work. The Windsor Transmission Plant was one of the oldest plants in Canada, having opened in 1920 to build engines and converted to an automatic transmission plant in 1963 as the popularity of such transmissions grew. But the plant was increasingly considered both "too antiquated" and "too expensive in terms of property taxes and utility rates" due to its "inner-city" location in the small city of Windsor (Vander Doelen and Turcotte 2008). The Windsor Transmission closure occurred despite a bevy of local and provincial incentives. "We pulled everything out of the hat over the last 18 months but if there's no product there's nothing to chase," said Ontario's minister of economic development. GM concurred: "Everybody was asking what they could do. But there was no product," said a spokesman (Vander Doelen 2008d).

Over the border in Michigan the automotive plant news was just as bad. In Pontiac, an industrial city thirty miles north of Detroit, GM has closed additional facilities, including the Pontiac East Assembly Plant, shedding an additional 1,100 jobs from the deeply troubled company. The loss of this plant in September 2009, whose opening in 1972 was a boon to the rapidly suburbanizing city, is a serious blow, estimated to cost the city a 20 percent drop in tax revenues in addition to the huge cost in reduced patronage and business for suppliers of the plant. Local businesses have felt a serious blow: "I used to make $120 in tips by noon, and today I haven't made $20," said an employee of a nearby restaurant soon after the plant closed. Employees of the plant felt similarly. "I don't know what to do," said a depressed worker. "It's still sinking in." The production of "light-duty trucks" formerly undertaken at the plant has been shifted to plants in Flint and Mexico (Martindale 2009). Local journalists were hopeful, imagining movie studios and other "new and equally vibrant" activities emerging from the wreckage of what was once by far the city's largest employer.

Amidst the ceaseless losses of the 2008–09 crisis, economic development officials were able to deliver a few pieces of better news. In Wixom, a distant exurb of Detroit, a Ford assembly plant that opened in 1957 just managed to make its fiftieth anniversary before closing in 2007, shedding 1,100 jobs in the process. Although Ford had comprised 12 percent of Wixom's tax base, the city was optimistic, stating that it was "achieving the transition to new economy status" typical of growing economies like Boston and the Bay Area. With the plant closed, the Michigan "State House New Economy and Quality of Life Committee" approved a tax credit package of $100 million over four years to attract three alternative energy companies, Xtreme Power, Clairvoyant Energy, and Oerlikon Solar, to convert the shuttered plant into a facility manufacturing electric batteries, solar panels, and wind turbines (*Tenders Info* 2009). Optimistically, the companies forecast the creation of 4,300 jobs, including suppliers. The refurbished plant expects to open in 2011 (McCauley Branstetter 2010). The $100 million tax credit, however, required a special vote of the Michigan Legislature as the state had already exhausted its $750 million budget for tax incentives for the year (Bunkley and Vlasic 2009).

Whereas new industries like green energy may reopen a fraction of Detroit-Windsor's shuttered automotive facilities, they are unlikely to recapture more than a small fraction of the jobs lost since 2000. Industrial retention or reopenings possess symbolic value, but they do not re-create the vibrant automotive economy that made Detroit and Windsor a metropolitan industrial powerhouse and that literally shaped much of the sprawling suburban landscape of the region. Several factors indicate that more trouble lies ahead within the coming years. The long-term health of the automobile industry is far from certain; despite the companies' new stability post-2009, the massive industrial economies of China and India have not yet entered the automobile market. The manufacturing of automobiles costing less than $3,000 in India indicates that further competition to the disadvantage of the American automotive industry is almost inevitable.

At the same time, the reuse of former automobile plants is likely to be difficult and expensive. Their locations, determined according to the economic and transportation calculus of fifty to seventy years ago, are no longer competitive for "new economy" uses. Wixom is located at the far urban fringe in an area with much open land, low costs, excellent accessibility, and a lack of deteriorated neighborhoods; most other locations lack one or more of these favorable characteristics. The least competitive facilities are those constructed prior to the Second World War along rail lines in relatively dense urban neighborhoods, such as those in Detroit and Dearborn. Despite the substantial restructuring of surrounding neighborhoods in those cities (Ryan and Campo 2007), the future of plants such as the Chrysler Jefferson East Plant and the GM Poletown Plant looks grim. Even GM admits that many of its facilities are "located in communities where there is very little interest in real estate markets" (Karoub and Runk 2010), making their future reuse problematic.

Many plants, despite their relatively recent construction in the mid- to late twentieth century, can be considered brownfields. "[These] are huge complexes that are going to take huge investments to get them back into productive use," note the authors of an Associated Press study (Karoub and Runk 2010). The same study cites an even more debilitating statistic: collectively, the Big Three have closed a total of 128 plants in the United States since 1980. Sixty percent of them have never been reopened, either remaining vacant or being demolished. The jobs picture is even direr: those 128 plants once employed 196,000 people. Today their replacements (the 40 percent of plants that have reopened) employ only 36,500 people (Karoub and Runk 2010). The overall picture is clear: automotive manufacturing is diminishing not only in Detroit-Windsor metropolitan area but nationwide and new jobs, if they are being created, are being created elsewhere.

THE CASINO INITIATIVE: REVITALIZATION THROUGH EXPLOITATION

In the face of this debilitating reality, a parallel development strategy is being pursued on an international scale to not only create jobs but to demonstrate the continued vitality and vibrancy of the Windsor-Detroit area. These new developments, housed in giant buildings employing thousands of people, have both economic and symbolic value, and their physical presence and local visibility is great. Unlike the suburban, physically unprepossessing automobile plants of the past, these new developments are located directly in the downtowns of Windsor and Detroit. The economic rationale for these developments is not manufacturing but leisure and tourism, an economy of consumption rather than production that is shaping the economic development strategies not only of Detroit and Windsor but of cities across North America and around the globe (Judd and Fainstein 1999, Judd 2003). These new developments, of course, are casinos.

The twenty-first-century construction of the new casinos that dot the downtowns of Detroit and Windsor has been undertaken in the same spirit of capitalist and cross-border competition that characterized the auto industry during the twentieth. The casino competition, however, is somewhat different in geographical scale and ambition than that of the automobile industry. Whereas the Big Three were, and to some extent remain, locked in a global competition to produce innovative products at the lowest cost, the casinos of Detroit and Windsor are unable to compete on a global level as leisure destinations. They therefore compete with their national peer cities, with each other, and with smaller-scale leisure and gambling facilities in their own suburbs. Like dozens of other states, provinces, and metropolitan areas across North America, the cities of Detroit and Windsor are now increasingly locked into a competition with each other to capture more leisure dollars than the other side of the border.

The first casino in Detroit-Windsor opened in 1995 along the Windsor waterfront. Originally titled the Windsor Grand Casino, with 700,000 square feet of space and a 200-room hotel, the casino immediately drew thousands of visitors per day, mostly from the Detroit side of the river (Ankeny 1998). Prior to the September 11, 2001 attacks, passage between Detroit and Windsor was quite easy, with only oral confirmation of identity generally proving sufficient to gain entry either to the United States or to Canada. The success of the Windsor grand demonstrated that the casino market of the region was strong and worth exploiting. In 1995 the Windsor casino demonstrated its success by grossing C$577.3 million. By the late 1990s, just before Detroit opened its own casinos, Windsor was grossing C$840 million per year (Rennie 2009). This immense profitability no doubt contributed to Windsor having the second-highest employment growth rate in Canada during the 1990s (*Windsor Essex* 2008: R-18).

The solo success of the Windsor casino was not long in attracting envy, and competition, from the American side of the river. In November 1996, Michigan voters approved "Proposal E," which authorized the construction of up to three casinos in downtown Detroit. Both the mayor and the city council supported the proposal because of the number of jobs that the casinos promised to bring. Industry experts predicted that as many as 11,300 jobs could be created, generating up to $1.5 billion in annual revenue for the casinos (*Crain's Detroit Business* 1996). The attraction of casinos for both local and state politicians, inured both to the relentless deterioration of Detroit and to the unhealthy regional automotive economy, is not hard to imagine.

Proposal E was designed to provide business opportunity for locally based organizations interested in opening casinos. Two such companies, 400 Monroe Associates and Atwater Entertainment, Inc., received two of the three casino licenses, with the third going to MGM Grand, Inc. of Las Vegas. All three companies were granted the right to open temporary casinos in existing buildings in the city while finalizing their "permanent" sites. The peculiar "temporary casino" strategy was linked to a misguided planning notion that "clustering" the casinos along the city's downtown waterfront, in a manner directly imitative of Windsor Grand's location across the river, would be a more successful development strategy (Serwach 1998). Ultimately the difficulty of acquiring land and the success of the casinos in their existing "temporary" locations led to the abandonment of this strategy in 2002.

The first of the "temporary" casinos to open was the MGM Grand. As proposed in 1997, the MGM casino project consisted of a

100,000-square-foot casino in addition to an 800-room hotel, restaurants, movie theaters, and even 70,000 square feet of convention and meeting space, all located along the eastern edge of State Route 10 leading to the Canadian border. By 2002, with the agreement to abandon the idea of the riverfront casino cluster, MGM committed to rebuilding its casino in a permanent location only one block north of its temporary one. The new casino was even larger, occupying several former city blocks with a sixteen-story hotel and a large shopping and restaurant complex. The total cost of this new casino was $800 million.

Perhaps the most distasteful episode in Detroit's casino construction was that of the Greektown Casino, financed in part by two local entrepreneurs with long-standing professional ties and substantial political influence. As a result, one of the casino contracts was earmarked for the Greektown project. The 'temporary' site was located in the midst of the busy Greektown restaurant row and adjacent to Interstate 75 in structures that were coincidentally purchased in anticipation of casino legalization. However, before the casino could open, both entrepreneurs were forced to sell their shares in the company as the result of a background investigation. The Greektown casino was therefore the last of the three casinos to open in late 2000. Like MGM, the Greektown casino also expanded, purchasing additional land for its own hotel, theater, and convention spaces.

The third and final Detroit casino was called the Motor City Casino. Motor City was owned by a local company but developed in partnership with Circus Circus Enterprises, Inc., again of Las Vegas. The Motor City casino was located only half a mile north of the MGM Grand, also along Route 10 freeway. Motor City actually reused a piece of Detroit's industrial fabric, occupying an abandoned Wonder Bread factory with 75,000 square feet of gambling space and several restaurants. Additions were made to the complex beginning in 2005 as the casino added hotel conference space and theater facilities.

The casinos worked their gambling magic. By late 2000, the two open Detroit casinos and the Windsor Grand were each generating over a million dollars in revenue daily. By 2008, the three casinos together had generated $1.36 billion for the year (Rennie 2009), in line with analysts' original forecasts of $1.6 billion per year, with up to $200 million of those revenues being returned to local, state, and provincial governments in taxes. On a regional level, Detroit and Windsor were successful, drawing visitors from as far away as Cleveland, according to one hotel operator. But analysts were skeptical, doubting the competitiveness of such a "cold-weather" region. As one drily noted, "This [Detroit-Windsor] isn't a destination market, it's a local market" (Slavin 2000).

Whereas the revitalization of downtown Detroit was an obvious aim of Michigan's casino acquiescence, another equally obvious aim, given the decision to site all three casinos along interstates leading to Canada, was the interception of dollars that might otherwise be spent across the border.

The 2001 crackdown on what was formerly the world's longest undefended border, while operating much to the detriment of automotive industries accustomed to 'just-in-time' delivery of parts to and from plants on either side, benefited the Detroit casinos. U.S. officials fearful of Canadian terrorists tightened entry regulations by requiring both birth and identity documentation as well as extensive questioning and even random searches of vehicles. Faced with delays at the border and the perception of a difficult crossing, many Americans simply stopped going to Windsor. By 2008, U.S. citizen journeys to Canada had dropped to the lowest numbers since 1972, dramatically depressing foreign visitation to the country as a whole (Quinn and Deslongchamps 2008).

The effects of this visitor depression on the Windsor casino were disastrous. Between 2000 and 2008, Windsor's gambling revenues dropped 60 percent, and employment declined in concert, dropping from 5,400 in 2001 to only 2,000 in 2009 (*The Economist* 2009; Vander Doelen 2009). In a desperate attempt to recapture visitors, the Windsor Casino was rebranded and rebuilt as "Caesar's Windsor," with the Ontario government investing over C$400 million (Vander Doelen 2008c) in a renovated structure with a twenty-seven-story hotel tower (the tallest structure in Windsor), a 5,000-seat "Colosseum," several new "upscale eateries," and a 100,000-square-foot convention space, second largest in Ontario (Glaser 2008, Macaluso 2008). With up to 80 percent of visitation dependent upon U.S. residents, Windsor began offering desperate incentives (Glaser 2009). In June 2008, the new Caesar's Windsor decided to offer $3,500 in tunnel fare to U.S. visitors on a first-come, first-served basis (Vander Doelen 2008b). By 2009, Caesar's was even offering application forms (Hall 2009) and free passport photos (McArthur 2009) to encourage U.S. visitation.

On the other side of the border, Detroit's reign atop the local casino food chain seemed impregnable, but nothing lasts forever. Elsewhere in Michigan, the Potawatomi tribe, after years of legal struggles, opened their own "FireKeepers" casino in 2009 less than two hours west of Detroit on Interstate 94 (Hepker 2009). Another casino in Gun Lake, Michigan, was also preparing to open in early 2011, creating additional competition for casinos already hampered by the post-2007 economic downturn (*Associated Press State & Local Wire* 2010; Daly 2009). Competing for a shrinking pool of money and a growing tide of unemployed, these new casinos seemed bound to create only more trouble for both Detroit's and Windsor's cars-to-casinos strategy.

CONCLUSION

This chapter can only begin to illustrate the complexity, challenges, and ultimate futility of attempting to arrest the decline of a globalized industry with significant local origins and history through the transferal of another globalized industry with no local origins or history whatsoever. Amidst the ebb and flow of global capital and global goods, the latest crisis (2008 onward) has dramatically reinforced how the Detroit-Windsor transnational metropolitan area, and even Michigan and western Ontario, are fading from the global scene, as they have been fading for decades. Detroit and Windsor's citizens may be global consumers, but their region's global contribution is dropping steadily as the Big Three automotive industries lose market share, shed employment and profitability, and ultimately risk and surrender their independent existence as corporations. In this de-globalizing region, the economic, social, and physical consequences of the death of the auto industry are painful to behold.

Public policy, whether at the local or provincial or state level, can do little to alter this bitter economic calculus. When federal subsidies are required to return a fraction of lost jobs, as in the case of Windsor's Essex Ford plant, the ultimate futility of the retention or attempted regeneration of automotive manufacturing-sector jobs becomes clear. The long-term manufacturing trend downward for the region seems irreversible, and when decline is steep and rapid, as it has been since 2000 and increasingly since 2008, little forward progress is possible.

The only viable alternative for this binational region dependent upon the dying automobile industry seems to be the new leisure economy of recreation and consumption. But in the very different and very competitive world of leisure, the "cold-weather" cities Detroit-Windsor cannot compete on a global or even on a national scale. With cheap air tickets and abundance choices, the local casinos will never be able to outmatch their better-equipped and longer-running competitors. The fundamentals of the leisure equation are unalterable: even had they ten or twenty casinos, Detroit and Windsor would likely be unable to compete with Atlantic City or Las Vegas. The region's new leisure economy of gambling has provided new jobs, buildings, and construction activity that has boosted the spirits of local leaders and citizens, provided thousands of service-sector jobs, and made a few headlines, but the cost, likely never to be tallied, has undoubtedly been high.

Lost amidst the gleeful revenue projections of Detroit-Windsor's four casinos is the dispiriting fact that a great amount of casino revenue has come from the accumulated savings of local residents. In their postindustrial decline, the region has essentially embarked upon a privately operated taxation strategy, with the major change being that the majority of the revenues generated has gone to casino corporations rather than to the government. Only in the most desperate economic circumstances would such a self-predatory strategy make sense. The recent casino expansions are a further sobering reminder that expansion is the only means for these facilities, in the absence of a growing economy, to compete. In 2010, the retreat of the automobile industry, and the incursion of the predatory casino industry, into Detroit and Windsor is doubtless far from over. How far the tides of globalization will ultimately retreat in this binational region and what further physical, social, and economic damage manufacturing will leave behind as it continues its inexorable decline remain to be seen.

NOTES

1. The term, as defined by Sassen (2001) and others, is predicated on modern postindustrial, finance-based economies and thus is not entirely applicable to the former industrial economies.

REFERENCES

Judd, Dennis, and Fainstein, Susan (1999). *The Tourist City*. New Haven, CT: Yale University Press.

Judd, Dennis (2003). *The Infrastructure of Play: Building the Tourist City*. Armonk, NY: M.E. Sharpe.

Rist, Gilbert (1997). *The History of Development: From Western Origins to Global Faith*. New York: Zed Books.

Ryan, Brent, and Daniel Campo (2007). "The (De)urbanization of Detroit: A History of the Automotive Industry, Economic Development, and Demolition." Presented at the 12th National Conference on Planning History, Portland, ME, October 25–28.

Sassen, Saskia (2001). *The Global City: New York, London, Tokyo. Second Edition*.Princeton, NJ: Princeton University Press.

Sugrue, Thomas (1996). *The Origins of the Urban Crisis: Race and Inequality in Postwar Detroit*. Princeton, NJ: Princeton University Press.

Thomas, June Manning (1997). *Redevelopment and Race: Planning a Finer City in Postwar Detroit*. Baltimore: Johns Hopkins University Press.

PERIODICALS AND NEWSPAPERS

Ankeny, Robert (1998). "Windsor Ups Casino Ante: Glitzy Mecca for Gamblers Set to Open." *Crain's Detroit Business*, July 20.

Associated Press Financial Wire (2010). "Chrysler Says It Reduced Incentives in 1Q," April 25.

Associated Press State & Local Wire (2010). "Gun Lake Tribe Announces $165M Loan for Casino," July 20.

Buckwalter Berlooz, Corry (2010). "Michigan Revs Up for Diversity: Cars are no Longer the Only Game in Town." *Planning* 76 (3): 10–15.

Bunkley, Nick, and Bill Vlasic (2009). "Looking to Green Energy to Drive Growth; Michigan Hopes to Offset Its Loss of Auto Jobs with Alternative Industries." *The International Herald Tribune*, September 11: 18.

Crain's Detroit Business (1996). "Week in Review: Election Results," November 11.

Daly, Pete (2009). "Casino Efforts Mixed." *Grand Rapids Business Journal*, September 14.

De Bono, Norman (2007). "A City on the Ropes." *London Free Press*, October 22.

DesRosiers Automotive Consultants, Inc. (2010). "Market Snapshot." http://www.desrosiers.ca, July 18.

Economist (2009). "The Humbling of Detroit North; Canada's Stalled Economy," August 1.

Glaser, Susan (2008). "All Hail the Newest Caesars." *Newhouse News Service,* September 23.

Glaser, Susan (2009). "New Passport Rule, Weak Economy Have Reduced Travel to Canada." *Plain Dealer Travel Editor,* July 26.

Guilford, Dave. (2010) "The New Detroit 3; What a Difference a Year Makes for Ford, Chapter 11 Twins." *Automotive News,* May 24.

Hall, Dave (2009). "Casinos, Bingos, Fear Loss of U.S. Customers." *Windsor Star,* May 12.

Hepker, Steven 2009). "Place Your Bets; FireKeepers Casino Opens Its Doors to Gamblers." *Jackson Citizen Patriot* (Michigan), August 6.

Karoub, Jeff, and David Runk (2010). "Cities Struggle to Find New Uses for Auto Plants." *The Associated Press,* January 11.

Keenan, Greg (2010). "U.S. Auto Sales Outpace Canada's: Year-over-Year Gains Outpace Those in Canada, Where Tepid 4-Percent Overall Rise Makes Strong Sales from Ford, Hyundai and Chrysler." *The Globe and Mail* (Canada), May 4.

Krisher, Tom (2010). "Ford Expected to Ride Sales Momentum, Good Vibes from Customers to First-Quarter Profit." *The Associated Press,* April 26.

Macaluso, Grace (2008). "Caesars Boosts Downtown Business." *Windsor Star,* July 15.

Martindale, Mike (2009). "GM Plant Closing May Cost Pontiac $10M." *The Detroit News,* June 2.

McArthur, Donald (2009). "U.S. Gamblers Get Passport Help." *Windsor Star,* March 26.

McCauley Branstetter, Nancy (2010). "Wixom Plant's Rebirth Offers Hope to Neighboring Leon's." *Oakland Press,* March 21.

McCracken, Jeffrey (2005). "Chrysler Expects Stable Work Force for Next 2 Years." *Detroit Free Press,* January 11.

Naughton, Keith, Jeff Green, and David Welch (2010). "Carmarkers in U.S. Are Running Once Again; GM Paid Off Debt; Ford Still Climbing." *Buffalo News,* April 25.

New York Times Topic (2010a)."General Motors." *New York Times.* Available at http://topics.nytimes.com/top/news/business/companies/general_motors_corporation/index.html; accessed on July 19.

New York Times Topic (2010b). "Chrysler LLC News." *New York Times.* Available at http://topics.nytimes.com/top/news/business/companies/ chrysler_llc/index.html; accessed on July 19.

Petry, Corinna (2010). "Ten Largest Automakers on Growth Path, Report Says." *American Metal Market,* April 12.

Quinn, Greg, and Alexandre Deslongchamps, Bloomberg News with files from Paula McCooey (2008). "Canada Has Fewest U.S. Visitors Since 1972: StatsCan." *Ottawa Citizen,* August 19.

Rennie, Gary (2009). "Casino Rebounds, OLGC Says; Caesars Sees Gains after Posting Poorest Year in its History." *Windsor Star,* January 16.

Serwach, Joseph (1998). "Agreements Paint Pictures of Wonders on Riverfront." *Crain's Detroit Business,* March 23: 32.

Schmidt, Doug (2009). "Nemak Closure Called a Blow to Workers, City." *Windsor Star,* February 14.

Shepardson, David (2009). "Bankruptcy Word Won't Go Away." *The Detroit News,* February 18.

Slavin, Al (2000). "In Casino War, Windsor Wins the Quarter." *Crain's Detroit Business,* October 30.

Targeted News Service (2010). "U.S. Labor Department Issues Final Exemption to Allow New Health Plan for Ford Motor Co. Retirees to Acquire Company Securities," March 25.

Tenders Info (2009) "United States: Ford, Granholm Discuss Plans for Energy Park," September 11.

Tenders Info (2010) "United States: GM Rides Cost Cuts, New Model Sales to 1Q Profit," May 18.

Tweh, Bowdeya (2010). "Ford CEO Touts Focus on Profit Drivers: Alan Mulally in Chicago Tuesday to Meet with Region Officials, Execs." *The Times* (Munster, Indiana), June 09.

Vander Doelen, Chris (2008a). "Feds Spur Ford Plant Rebirth $80 Million Pledge for Essex Engine to Be Announced by PM Today." *Windsor Star,* September 03.

Vander Doelen, Chris (2008b). "The Big Gamble; World-Renowned Gaming Company Has High Hopes for City. But Will It Work?" *Windsor Star,* June 17.

Vander Doelen, Chris (2008c). "Caesars Windsor 'Breathtaking'; Windsor's Two Cabinet Ministers Issued Similar One-Word Reviews Friday after Their First Glimpse of the New Caesars Windsor: 'Wow' and 'Breathtaking.' " *Windsor Star,* June 13.

Vander Doelen, Chris (2008d). "'No Product,' No Plant, Despite Leaders' Efforts." *Canwest News Service,* May 12.

Vander Doelen, Chris (2009). "Casino Rebound a Bad Bet." *Windsor Star,* October 3.

Vander Doelen, Chris, and Rebecca Turcotte (2008). "Windsor GM Plant Closure in 2010 Will Leave 1,400 Jobless; CAW's Hargrove 'Frustrated, Angry' at Lack of Projects." *Canwest News Service,* May 13.

PROFESSIONAL LITERATURE

Center for Automotive Research (2008a). *Beyond the Big Leave: The Future of US Automotive Human Resources* (2008). Center for Automotive Research. Available at http://www.cargroup.org; accessed July 13, 2010.

Center for Automotive Research (2008b). *Southwest Ontario Automotive Outlook* (2008). Center for Automotive Research. Available at http://www.cargroup.org/documents; accessed September 28.

Gomes (2008). *Economic & Auto Industry Outlook: 2009 Steering through Turbulence.* Carlos Gomes, Senior Economist, Industry Research, The Scotiabank Group, Toronto. Available at http://www.choosewindsoressex.com; accessed September 28, 2010.

Ford Motor Company 2002 Annual Report (2002). Dearborn, MI: Ford Motor Company.

Hoover (2010). *Ford Motor Company: Historical Financials.* Hoover's Company Records. Available at http://premium.hoovers.com/subscribe/co/fin/history.xhtml?ID=ffffrfhskffrycckcj; accessed July 20, 2010.

Fortune (2008). Fortune 500 Companies—Archived List of Best Companies from 1995. *Fortune* Magazine. Available at http://money.cnn.com/magazines/fortune/fortune500_archive/full/1955/.

OICA Statistics Committee May 2001.*World Motor Vehicle Production by Manufacturer—World Ranking 2000* (2010). Available at http://oica.net/category/production-statistics/2000-statistics/; accessed July 26, 2010.

Orbis (2010a). *Orbis Company Report—Ford Motor Company.* Available at https://orbis.bvdep.com/version-2010623/cgi/template.dll?product=13&user=ipaddress; accessed July 21, 2010.

Orbis (2010b). *Orbis Company Report—Motors Liquidation Company*. Available at https://orbis.bvdep.com/version-2010623/cgi/template.dll?product=13&user=ipaddress; accessed July 21, 2010.

Orbis (2010c). *Orbis—DaimlerChrysler AG Form 20-F Filed February 20, 2002*. Available at https://orbis.bvdep.com; accessed July 22, 2010.

Orbis (2010d). *Orbis—DaimlerChrysler AG Form 20-F Filed February 28, 2005*. Available at https://orbis.bvdep.com; accessed July 22, 2010.

Orbis (2010e). *Orbis—DaimlerChrysler AG Form 20-F Filed February 27, 2008*. Available at https://orbis.bvdep.com; accessed July 22, 2010.

Statistics Canada (2010). *Table 6–1 Labour Force Characteristics—by Province and Economic Region, Unadjusted for Seasonality, 3 Month Moving Average Ending in May 2009 and May 2010*. Statistics Canada, 2010. Available at http://www.statcan.gc.ca/pub/71–001-x/2010005/t021-eng.htm; accessed July 19, 2010.

U.S. Department of Labor Statistics (2010a). *State and Area Employment, Hours, and Earnings—Michigan, Motor Vehicle Parts Manufacturing*. Bureau of Labor Statistics, 2010. Available at http://data.bls.gov/PDQ/servlet/SurveyOutputServlet?series_id=SMU26000003133630001&data_tool=XGtable; accessed July 19, 2010.

U.S. Department of Labor Statistics (2010b). *State and Area Employment, Hours, and Earnings—Michigan, Motor Vehicle Manufacturing*. Available at http://data.bls.gov/PDQ/servlet/SurveyOutputServlet?series_id=SMU26000003133610001&data_tool=XGtable; accessed July 19, 2010.

Wards Automotive Group (2010). *U.S. Total Vehicle Sales Market Share by Company, 1961–2009*. Available at http://wardsauto.com/keydata/historical/UsaSa28summary/; accessed July 22, 2010.

Windsor Essex (2008) Windsor–Essex County Development Commission 2006. Available at http://www.choosewindsoressex.com; accessed September 28.

Zacks Equity Research (2010). *Auto Industry Outlook and Review*—April 8. Available at http://www.zacks.com/stock/news/32667/Auto+Industry+Outlook+and+Review+%96+, accessed July 22, 2010.

6 From a Fishing Village via an Instant City to a Secondary Global City

The "Miracle" and Growth Pains of Shenzhen Special Economic Zone in China

Xiangming Chen and Tomás de'Medici

> Shenzhen and Dubai may have outstripped Paris and New York as civic models. But can an instant city ever feel like the real thing? Built at phenomenal speeds, these generic or instant cities, as they have been called, have no recognizable center, no single identity. It is sometimes hard to think of them as cities at all . . . (Ouroussoff 2008: 70, 72)

INTRODUCTION

If not for the dazzling pace at which both Shenzhen exploded from a tiny fishing village on the Hong Kong border and Dubai rose from the desert in the Middle East, it would be hard to fathom that a *New York Times* architectural critic would mention them in the same breath as Paris and New York. But both Shenzhen and Dubai have risen at a dramatic pace, and each has become known as an "instant city," albeit with very different histories and different scales, forms, and functions, respectively.[1] Leaving the examination of Dubai to more qualified scholars (see Kanna 2011 and in this volume), we examine Shenzhen's miraculous growth over the span of three decades (1979–2009) to what may be regarded as a secondary global city (below Shanghai and Beijing), emphasizing the growth pains associated with its transition as an instant city.

Shenzhen is coming of age. In 1979, China's first special economic zone (SEZ) was set up in Shenzhen, a small fishing town bordering Hong Kong, for attracting overseas investment and initiating domestic reform. Shenzhen was an unexpected product of closed China with its centrally planned economy. Its progress was so rapid that it became, in some ways, a victim of its own success, in part because of internal and external constraints on its current and future health.

Shenzhen is confronting cumulative economic, social, and political challenges stemming from its superfast growth—it is China's largest city of migrants and is particularly vulnerable to the effects of the global economic crisis. A retrospective and prospective study of Shenzhen will reveal the complex lessons of a city that is touted for its growth rate, glitzy skyscrapers, and gross economic rankings among Chinese and world cities.

Shenzhen offers many insights into an array of contentious processes and outcomes that define it as an "instant city."

We first establish Shenzhen as China's fastest growing and first contemporary "instant city" one that differs considerably from seemingly similar cities in Western history. Next, we identify three salient features of Shenzhen that reflect both the progress and pain of its rapid economic ascent: its primary emergence as a dynamic SEZ; its evolution into a full-fledged industrial city; and its growth pains in transitioning to a more global city. This approach differs from a recent comprehensive review of China's SEZs, especially Shenzhen (see Yeung, Lee, and Kee 2009), through a critical evaluation of Shenzhen's speedy successes and subsequent challenges. By highlighting industrial transition, migrant labor, and governance reform, we will shed light on the broader implications of the Shenzhen model for development zones and city-building in China (also see Liu and Chen, this volume) and the broader lessons about the challenges and opportunities facing secondary cities that aspire to rise globally.

SHENZHEN AS AN "INSTANT CITY"

In terms of a city *ex nihilo* (out of nothing) (Braudel 1992), Shenzhen qualifies as an "instant city," but it has boomed on a larger scale, at a faster speed, and, most importantly, in a distinctive fashion. Shenzhen broke the record of constructing a skyscraper at an average of two and a half days per floor in the 1980s and continued to change its "skin" or cityscape in both wrenching and fluid ways (Ratti 2003). Shenzhen was once depicted as a "city that sprung up when someone just added water."[2] That "water and fertilizer" came largely from a powerful state that provided favorable policies and rapid buildup of large-scale infrastructure (see Liu and Chen in this volume on the similar role of the state in launching Tianjin Binhai New Area).

Shenzhen is not merely an instant city. It has risen higher and further along the hierarchy of world or global cities, to the point where it can be regarded as a secondary global city. Shenzhen merits secondary global city status assuming Shanghai and Beijing are either already at or close to the top tier of the global city hierarchy with their greater demographic and economic rankings. Among 150 metropolitan areas globally, Shenzhen ranked first in income and employment growth from 1993 to 2007, and held the number four spot through the recession period of 2007–10, behind Beijing, Guangzhou, and Shanghai (Brookings Institution and London School of Economics 2010). By hosting the 26th Summer Universiade Games in August 2011, Shenzhen has projected the message of having arrived on the world stage, blasting the persistent glamorous TV promotion about the city on CNN even after the games ended. To understand the transition from an instant city to a secondary global city, we employ the concept of "coming of age" to define Shenzhen's developmental pathways and prospects. We undertake a broad analysis of three crucial dimensions of Shenzhen that have shaped its existence and identity and may well constrain its future: economic growth, migrant labor, and political governance. This coming of age reveals the cumulative challenges—"growth pains"—of being an instant city that is trying to move beyond its origins.

CHINA'S FIRST SPECIAL ECONOMIC ZONE

Shenzhen's renowned success was tied to its status as the first SEZ created at the beginning of China's historic turn to modernization and internationalization. In 1978, China was poised to enter a new era after emerging from the "dark decade" (1966–76) of the Cultural Revolution. The post-Mao leadership under Deng Xiaoping—who had returned to power after the Cultural Revolution—introduced the "four modernizations" in agriculture, industry, technology, and defense. Included in this development strategy was the Joint Venture Law. Passed in 1979, this law allowed overseas enterprises to establish equity joint ventures with Chinese companies and directly exposed state-owned enterprises to the financial and managerial influences of foreign investors. Besides immediately unleashing major changes at the organizational level, this also affected larger institutional arrangements as corporate financing through state-controlled banks increased. This could be seen by China's emergence as the most popular destination among developing countries for foreign direct investment (FDI).

It is widely known that Deng Xiaoping was the principal decision maker of the establishment of special economic zones. As early as April 1979, after listening to the reports given by the main leader of the Guangdong Provincial Party Committee, Deng remarked: "We can mark out a piece of land and call it a special zone."[3] This led to the promulgation of "The Regulation on Special Economic Zones" in 1980, which defined a "special economic zone" as "an area where enterprises are treated more preferentially than in other areas in relation to such matters as the tax rate and the scope of operations in order to attract foreign capital and advanced technology for modernization" (Chen 1994, 1995). The SEZs furthered development of public utilities and fostered a receptive investment climate. They offered preferential treatment to foreign firms seeking advantages in land use, rent agreements, and tax holidays. The SEZs were designed to help major exporters of manufactured goods engage global markets. The concept and intended practices of SEZs were not entirely new, as they bore a resemblance to the export processing zones (EPZs) that Taiwan, South Korea and other countries had been using since the 1960s to drive their export-oriented industrialization efforts (Chen 1994, 1995; Farole, 2011; Sit 1988). However, SEZs were a pioneering concept for China, with its firmly entrenched socialist political economy that lingered into the 1980s. As the first SEZ, Shenzhen was the foundational locality in China to be cordoned off to experiment with economic reform and market activities—a sort of testing ground for capitalist practices. Thus began China's gradual move from centralized planning and isolationist policy to a more market-friendly economy. It was a bold strategy tempered by cautious tactical steps. Although an unprecedented move to introduce market practice and foreign investment to postreform China, it was implemented with the state's unfettered ability to control the experiment in a confined environment and restrict negative effects. This coupling of audacity and caution enabled successful replication and risk reduction from the genesis of China's reform when major success was dubious and the risks of failure were perceived as high (Chen and de'Medici 2010; Farole 2011).

FROM TORRENT GROWTH TO TOUGH TRANSITION

Shenzhen has incomparably maximized the first-mover advantage in achieving instant rapid growth. This is not surprising given Shenzhen's pregrowth backwardness and geographic adjacency to Hong Kong and its plentiful capital. Like the rest of China, pre-SEZ Shenzhen had a history of economic stasis that lasted until reform sparked and sustained its remarkable growth. Shenzhen's power and status aside, the first mover advantages raise questions about its long-term strength. We provide an analytical peek into both the predictable origin and trajectory of an instantly booming city and the protracted challenges it faces as it matures.

Shenzhen exploded from a small fishing village—a tiny outpost at the far southern edge of the expansive Chinese mainland bordering Hong Kong—into the giant industrial metropolitan region of today. Even in the mid-1980s, when the senior author of this chapter first traveled to Shenzhen to conduct research, this young SEZ had relatively few paved roads and a barren landscape with relatively little construction of residential buildings and commercial service facilities. Instead, many new factories filled the vista and cityscape, heralding in earnest the emergence of a large industrial city.

Starting with a small population of about thirty thousand people in 1979, the population of Shenzhen rose to 8.9 million in 2009 (Shenzhen Statistical Bureau 2010). The true demography of Shenzhen may reveal an additional population of perhaps another four million or so migrant laborers who lived in the city temporarily and were not counted in the official tally; adding them would raise the total de facto population to approximately 12 to 13 million (Shen 2005; You 2008). Regardless of the tricky definition of urban population in China (see Shen 2005), there may be a 4:1 ratio of temporary population (6.5 million long-term temporary residents plus four million or so short-term migrants for a combined ten million) to a legally permanent population of 2.4 million or those with official household registration (*hukou*) in 2009 (see Figure 6.1). Whereas the line for the smaller *hukou* population is quite smooth and reflects strict government control over its size, the larger temporary population grew faster and fluctuated more. This demographic composition is quite unusual among China's largest and economically most important cities and poses challenges from this huge migratory workforce.

Shenzhen's annual average growth of 30 percent since 1979 was globally unequalled and it had a GDP of about $120 billion in 2009, ranking among the top five among all Chinese cities. Its gross domestic project (GDP) per capita in 2009 reached almost $14,000, the highest among all Chinese cities (see Figure 6.2),[4] with its GDP per capita projected to reach $20,000 by 2020, which will be another first for China.[5] Shenzhen industrialized at a greater rate than any other Chinese city between 1979 and 2009; its industrial share of the GDP rose from 20 percent in 1979 to 46.7 percent in 2009 while its agricultural share dropped from 30 percent to less than 1 percent (Shenzhen Statistical Bureau 2010). Shenzhen has ranked as China's number one export city in per capita terms for almost twenty years in a row. Thirty years ago, Shenzhen barely had a serviceable road linking its center to the outside world; it developed into an excellent logistics hub, boasting the world's fourth largest container port behind only those in Singapore, Shanghai, and Hong Kong, with its growth coming at the heavy expense of Hong Kong.

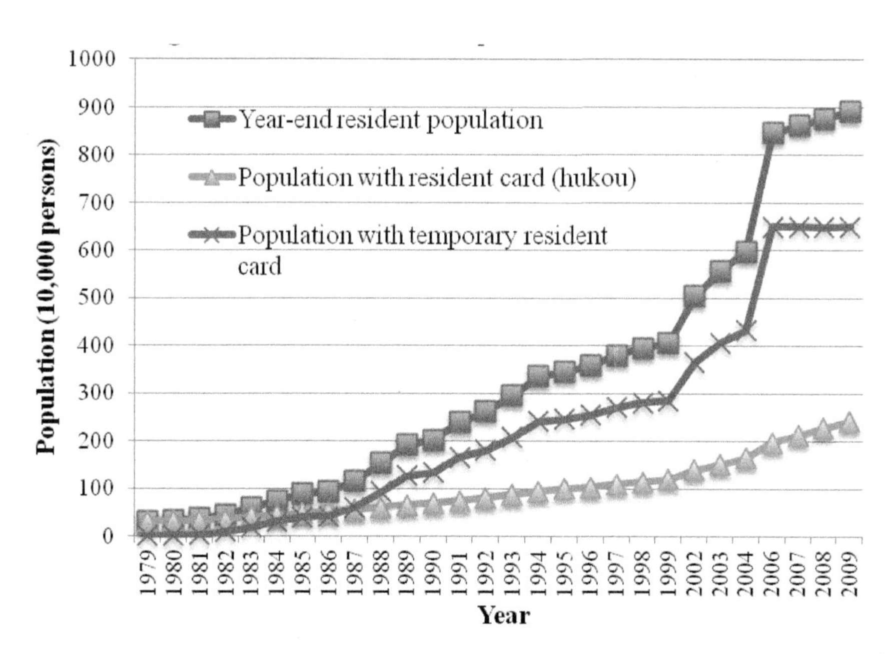

Figure 6.1 Shenzhen's population growth, 1979–2009.
Source: Graphed from Shenzhen Statistical Bureau (2010).

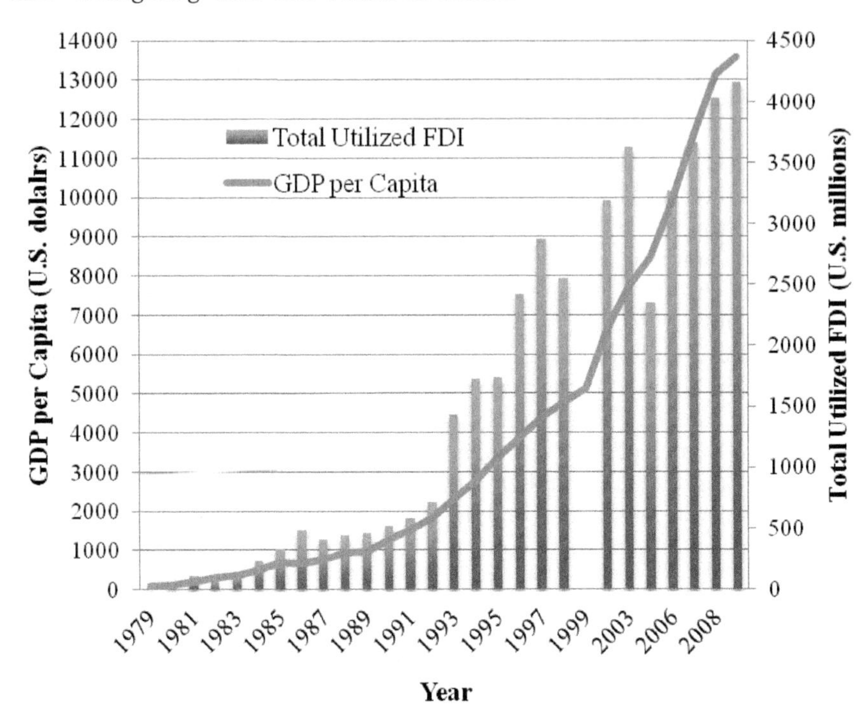

Figure 6.2 Shenzhen's GDP per capita and utilized foreign direct investment (FDI), 1979–2009.
Source: Graphed from Shenzhen Statistical Bureau (2010).
Note: Data for total utilized FDI in 1999 are missing.

First-Mover Advantages

Shenzhen would never have experienced such explosive demographic, industrial, and infrastructure growth without its first-mover advantages working quickly and intensively. Shenzhen became a living laboratory for new economic principles and practices that were alien to the socialist planned political economy prior to 1980. By offering lower taxes at 15 percent as opposed to 33 percent outside its boundaries, Shenzhen became the first and foremost destination for an initial wave of foreign investment. By March 1981, total foreign investment in Shenzhen had grown to $400 million (Nishitateno 1983: 176), far exceeding any other single locale in China. From 1983 to 1985, FDI in Shenzhen rose to about 75 percent annually. By the mid-1980s, more than 52 percent of all the equity joint ventures in China were located in Shenzhen and the three other SEZs (Zhuhai, Shantou, and Xiamen). The four zones accounted for 34.6 percent of the total pledged foreign investment in such projects, with Shenzhen getting the lion's share (Shenzhen Statistical Bureau 1985). Shenzhen's first-mover advantages got a big early lift from its proximity to Hong Kong. Given Hong Kong's historical economic relationships with international markets, it was well prepared to provide Shenzhen with an entry point to global commerce. This allowed for low-cost production in Shenzhen, where land and labor were cheap; shipping of raw materials and products to international markets, which were low-cost and convenient; easy supervision of the production process with excellent quality control; and streamlining coordination with Hong Kong headquarters. The result was the efficient and expeditious export of inexpensive goods to international markets through Hong Kong. From 1986 to 1993, Hong Kong accounted for 66 percent of the overall FDI in Shenzhen and 64.2 percent of the cumulative foreign investment (Liang 1999: 119).

Whereas geography reinforced Shenzhen's first-mover advantages, it did not sustain them later as Shenzhen's SEZ status gradually became less exclusive—its success prompted the government to extend the stimuli of FDI to other cities and locales across China. Shenzhen lost some of its first-mover monopoly in 1984 when the central government designated fourteen other coastal cities as economic development zones, each of which became a competitor with Shenzhen and with one another for FDI and fast growth. The zone set up by the northern coastal city of Tianjin in 1984 laid the foundation for launching the TBNA as the third coming of Shenzhen in the mid-2000s (see the next chapter by Liu and Chen).

The second coming of Shenzhen occurred in 1992 when Pudong, Shanghai, took off after Deng Xiaoping paid his return visit to Shenzhen, eight years after his first inspection tour in 1984. During his 1992 visit, Deng expressed his satisfaction with and confidence in the rapid development of Shenzhen as a SEZ and reconfirmed its importance as a model for the rest of China. Deng's visit provided an immediate boost to FDI in Shenzhen, which rose from $250 million in 1992 to $497 million in 1993. In 2009, the amount of utilized FDI in Shenzhen reached $4.2 billion (see Figure 6.2). This FDI indicator, which had already been sunk and put to productive use, was at its historical peak for this relatively young city.

Shenzhen's Growth Pains

The rise of Shenzhen in comparison to other Chinese cities was not without foreshadowing, given the decade-long head start that stemmed from its SEZ designation. Its first-mover advantage was magnified by its propinquity to Hong Kong. However, the glorious early years of Shenzhen's development did not last forever. Shenzhen experienced the inevitable growing pains of economic growth and attendant developmental adjustments. It ran into the constraints of limited land, shortage of energy and water, demographic pressures, and environmental contamination. The city's GDP per square kilometer, for instance, was only about 10 percent of Hong Kong's. Its water use efficiency measured by consumption per 10,000 Yuan of GDP was less than a third of that of Japan.[6] Even if the unofficial, estimated gross population total of 14 million in 2009 might be high, Shenzhen's official tally of 8.9 million—of which less than one in four people were living in registered households (Figure 6.1)—was a demographic issue that had to be factored into any comprehensive development strategy.

These development bottlenecks have been met with purposeful policy responses. Since 2005, the Shenzhen government has increased restrictions on land approval, raised the minimum wage to the highest among large Chinese cities ($125 a month), and elevated environmental standards by banning particularly polluting industries including dying, papermaking, tanning, and electric-coating. The inevitable rise in both land and labor costs due to rapid growth, coupled with both reactive and proactive government policies, led to some capital exodus and elevated the importance of timely and successful industrial restructuring in the affected sectors (Hu 2007).

Despite their initial catalyzing effects, tax breaks are no longer sufficient to induce businesses to maintain operations in Shenzhen. For example, Taiwanese-based Guangdong Nai Li Shoes has moved its factory to nearby Dongguan to secure cheaper land and labor. "We had tax privileges for five years," said owner Mr. Li. "Then we realized that Shenzhen was getting so expensive, so we have moved our 30,000 square meter factory with 4,000 workers to Dongguan."[7] The Shenzhen-headquartered Measurement Specialties, which makes sensors for installation into devices ranging from cars to pacemakers, has also left. Having recently acquired a small Chinese company in the Chengdu, Sichuan province, Measurement Specialties has begun planning to move some noncore, labor-intensive operations to Chengdu, where labor cost is about 20 percent lower than in Shenzhen.[8] Even Shenzhen's home-grown powerhouse Huawei—China's largest telecommunications company—began construction of a production facility on a 500,000-square-meter technology-park site in Dongguan. Huawei, which employs 22,000 people worldwide (including around 1,000 in Bangalore, India), maintains about 4,000 workers in Shenzhen and has invested $588 million in the Dongguan facility.[9]

In 2000, Ping An Insurance—which originated in Shenzhen in 1988 and became a major national company—began to relocate its core business units to Shanghai, and in September 2002, Ping An announced the construction of the $3 billion Ping An Financial Tower in Pudong, Shanghai. This was a substantive and symbolic shift of its corporate gravity away from Shenzhen and towards Mainland China's premier financial center. By the end of 2002, China Merchants Bank, which was founded in Shenzhen in 1987 as China's first shareholding commercial bank, moved its credit card center to Shanghai, where it would operate its nationwide credit-card business. Even Wal-Mart's China Procurement Center, once ensconced in Shenzhen, moved to Shanghai, dampening Shenzhen's aspirations to become an international logistics center.

With both homegrown and multinational companies leaving Shenzhen, a Shenzhen-based reporter for the BBS forum of the official *People's Daily* pondered "whether Shenzhen is being cast away, and by whom?"[10] Perhaps a mere rhetorical question, as Shenzhen—a key city with top performance indicators—cannot be "cast away." Nevertheless, the reporter's controversial question, published on November 17, 2002, marked a turning point in Shenzhen's coming of age and triggered new thinking and planning for the city's future that reflected and reinforced its pioneering role in economic reform and development over the previous two decades. By building up the local economy on a clean slate of export-oriented light industries and then moving toward high-tech industries, Shenzhen almost skipped the standard stage of heavy industrialization that many of China's other manufacturing centers experienced, distinguishing its path-dependent trajectory. While highly successful, it created imbalances and left a large void in Shenzhen's industrial foundation.

In response, Shenzhen promoted the growth of some heavy industries, which grew at 16.5 percent in 2007, faster than the 14.7 percent GDP growth (Wang 2008). This policy tilt was accompanied by the acceptance and encouragement of labor-intensive processing factories leaving Shenzhen for interior cities. In the first half of 2007, the six district governments of Shenzhen, working with some industrial associations and interior municipal governments, established twenty "enterprise transfer parks" to facilitate the relocation of less competitive factories to cheaper locations.[11]

Transcending its earlier successes in labor-intensive assembling and manufacturing, Shenzhen tried to establish itself as China's top-ranked city in a number of new national and international brand-name products. Shenzhen became adept in developing the capacity to design products instead of merely manufacturing them. In June 2008, the national government approved Shenzhen to be the first experimental city for becoming an innovation center.[12] With a vibrant graphic and industrial design industry encompassing more than 6,000 design companies and 60,000 designers, UNESCO named Shenzhen a "City of Design" in late 2008.[13]

Given its entrenched industrial history and the intense competition posed by more competitive cities like Shanghai for developing the service sector, which surpassed manufacturing and accounted for over 50 percent of GDP by 2009, Shenzhen may still find self-reinvention formidable. The maturation after thirty years of neck-breaking growth has indeed presented challenges to Shenzhen's future prosperity. However, with prudent planning, these challenges are also opportunities for an "instant city" to rebalance itself as a more permanent hub for sustainable development toward more global prominence as a stable secondary city even if it is destined, in the short run, to be economically subordinate to Shanghai and Beijing.

CHINA'S LARGEST CITY OF MIGRANTS

Shenzhen is China's largest city of migrants. Moreover, it has an extremely young population, averaging just less than thirty years of age (Ratti 2003).

Since Shenzhen's takeoff around 1980, millions of migrants have moved there from the rest of China to seek better economic opportunities (Chen 1987). The rare and glamorous success stories notwithstanding, most migrants have labored quietly to make a living. Whereas some have returned to their hometowns outside the city, most have stayed for varying periods of time and some have permanently set roots in Shenzhen and obtained legal resident status. Regardless of their individual gains and losses, these millions of migrants have transformed Shenzhen into an "instant city" of wealth and prosperity.

Migrants Rich, Poor, and In-Between

The fortune of earlier migrants to Shenzhen reflects the city's appeal as a newly opened frontier, rebelling against the mobility that Maoist China violently suppressed. The more calculated migrants retained their previous household registrations, as a precaution should they need to return home, whereas others risked everything. The migrants' stories are that of a boomtown that has allowed some to prosper and many to languish in a fiercely competitive environment.

Some migrants, through hard work and good fortune, have found themselves titans of a post-Maoist China with personal wealth in millions of U.S. dollars. However, most of the migrants who came seeking both quick money and a short stay found themselves earning too much to justify returning home, but too little to actually have the freedom to do so. Millions of workers have become de facto residents of the city. In the early years of the SEZ, it was reported that Shenzhen's policemen would put up their own street stands to sell merchandise after they finished their daily duties (Wang 2008). A growing number of migrants have risen to middle-class status through years of hard work and educational mobility. The stories of many of Shenzhen's migrants are reminiscent of the experiences of immigrants who arrived in booming American industrial cities in the nineteenth and early twentieth centuries.

The Misfortune of Outsiders

For the majority of the migrants to and in Shenzhen, success has proved far more elusive. They are not accorded household registration and its attendant privileges. This hinders social standing and access to affordable education, as well as subsequent professional credentials. Without low-cost access to education, they have to pay a large sponsoring fee for their children to be able to attend local schools. Compounding these barriers are high health-care costs. Migrant nurses are paid at most half the wages allocated to legal resident colleagues. Contrary to permanent residents, who can easily apply for and procure travel permits to Hong Kong, migrant workers must return to their hometowns to apply for the same privilege. The misfortune of migrants also exists in large inland cities like Kunming, where local taxi drivers perceive them as undesirable (see Notar, this volume).

Facing these and other discriminatory practices, migrant workers feel they are second-class citizens. For many, the longing for the real home manifests itself at Chinese New Year when Shenzhen is transformed into a ghost town as millions of migrants return home for the extended holiday. Whereas many migrants have lived in Shenzhen for some time, their weaker attachment to the city, coupled with their remittances reducing local consumption, has an aggregate negative effect of lowering local consumption. Shenzhen's retail sales in 2003 were half of Guangzhou's, even though the former average per capita income was higher than the latter. This prompted some retailers to set up their South China headquarters and flagship stores in Guangzhou rather than in Shenzhen (You 2008).

The Challenge of Incorporating Migrants

Addressing the current needs and long-term prospects for millions of migrant workers in Shenzhen is a major challenge: it bears the weight of sustained economic growth, accumulated wealth, and unwillingness to accept migrants as true local citizens. A former party secretary of Shenzhen Municipality advocated discontinuing the phrase "peasant workers" and replacing it with "outside workers." This nonetheless carried the connotation of migrant workers coming from the outside. A former mayor of Shenzhen expressed concern that if temporary workers are without personal urban identification, they are more inclined to leave, which would be particularly detrimental at a time when they were needed for industrial upgrading (Zhang 2007). Whereas Shenzhen used to attract about three hundred thousand college graduates every year, that number dropped by about half in recent years—which could be partly attributed to accelerated growth, more job opportunities, and higher pay in the Yangtze River Delta around Shanghai (Chen 2007).

Of late, the Shenzhen government has introduced several policy measures, some of which are the first in China (mirroring its lead in initiating other reforms). In May 2006, the Shenzhen government introduced a medical-insurance program for migrant workers mandating employers to pay one U.S. dollar with employees contributing fifty cents a month to increase access to medical treatment. The government labor bureau and lawyers persuaded some evasive factory owners to return millions of dollars in back pay to migrant workers.

This is now more arduous, as the recent global recession has prompted some business owners to shut down factories, sometimes overnight, and flee China to avoid paying both suppliers and workers. Worse, as rural migrants often live in huge factory dormitories, these workers have no place to live if the bosses flee. This has recently prompted the Guangdong Lawyers Association to call for legislation that will allow criminal prosecution of bosses who flee factories (they currently only face civil penalties) (Roberts 2009). In a city built by millions of migrant workers whose meager livelihood pivots directly on the vicissitude of the global economy, a severe crisis has engendered a major labor management and protection challenge that tests the local government's will and ability to respond effectively.

On August 1, 2008, the Shenzhen government took the biggest step thus far by beginning to issue the "permanent residence card"[14] to temporary migrant workers (which can help make the two lower trend lines in Figure 6.1 converge). In principle, this new card will allow migrant workers to enjoy the same housing, medical, educational, and pension benefits as those with *hukou* or household registration, despite their legal distinctions. Shenzhen planned to issue five million such cards by the end of 2008 and widen its coverage to over 90 percent of the migrant labor force by the end of 2009. Some migrants who have been in Shenzhen for a few years have already applied for and received the card. It is unclear if this will be sufficient to bridge the gap between the small proportion of Shenzhen residents with *hukou* and the majority migrant population that are expected to receive the permanent residence card. One of its immediate effects will be to blur the distinction between the "local and legal" resident with *hukou* and the "outside" migrant and thus erode the former's privileges. But if the new card is not *hukou*, it will test the government's desire and ability to make this program work to the point where the formal and informal distinctions between the permanent residence card and *hukou* will really disappear.

PIONEERING AND LEARNING IN GOVERNANCE

Shenzhen broke new ground in utilizing foreign investment and bank loans for constructing infrastructure projects. It adopted tendering where project design, construction, and furnishing would all be subject to competitive bids as a way of reducing risks and improving the efficiency of capital construction. This practice contributed to the "Shenzhen speed" of growth in ways other than being the "historical first." Shenzhen blazed the trail in introducing contractual employment through the first set of Chinese-foreign joint ventures, leading to early reforms of traditional lifetime employment, egalitarian wages, and insurance that was prevalent in state-owned enterprises. Furthermore, Shenzhen started the first foreign-exchange swap center in China, allowing some degree of openness, competition, standardization, and fluctuation based on market principles. Shenzhen also broke new ground by using auctions to transfer the use or leasing right to state-owned land.

These successful policy experiments in Shenzhen were quickly adopted in wider national practice in the 1980s. This process was a spatial extension of "better (albeit not best) practices" from Shenzhen to the rest of China. The transfer of land-use rights through auctions was crucial to the rapid buildup of the Pudong New Area of Shanghai in the early 1990s (see Chen 2009). This demonstrated that large infrastructure projects, like highways, could be built through a bidding process. Once it became replicable and repeated in other Chinese cities, it galvanized the rapid spatial transformation. First tested in Shenzhen, contractual employment spread quickly and widely to other cities and became an enduring element of China's new labor market.

Stretching the Reform and Governance Frontier

The central government recognized that Shenzhen was less tainted with strong control and deep economic planning, and thus placed it on an uncharted course for developing new and innovative governance reforms. These reforms took on a more political or institutional focus in place of earlier economic reforms to sustain growth. Like Deng Xiaoping and Jiang Zemin, who had blessed Shenzhen personally with their nationally reported visits, Hu Jintao, the current Communist Party chief, visited Shenzhen in April 2003 and accentuated its historical position as China's "open window and experimental ground" that could and should stay ahead of the rest of the country.[15]

Political reform has focused on the improvement of governance capacity through a comprehensive restructuring of the municipal government from an omnipresent one to a more limited one, from a regulator/administrator to a service provider and from a power holder and influence wielder to one with a civil service identity and accountability. This involves separating the government's functions of decision making, implementation, and monitoring, which aims to reduce the governmental tendency to "overreach, neglect, and misjudge" (Li 2008). The essence and scope of the reform reflects a serious effort to move into new political territories that carry far-reaching policy implications.

The recent political reform also includes some specific but ambitious measures. In May 2008, the Shenzhen government unveiled a plan to institute a multicandidate mayoral election and to introduce the Hong Kong model of clean and corruption-free government in three years. This would allow self-recommended and party-nominated candidates to run limited campaigns, which would take place first at the district or subcity level, where candidates could give open speeches and engage in public debates.[16] In November 2008, the party secretary of Shenzhen introduced a process through which the municipal government would move from the current "two-level government and four-level governance" structure to a new "one-level government and three-level governance" model. This envisioned the elimination of the district government's structure (which corresponds to that of the higher municipal government) and the creation of differentiated and simplified governance at the municipal, district, and street-office levels.[17] The goal was to lighten the top-heavy and top-down approach to governing and replace it with a more limited governance setup more attuned to community interests.

Governing a Transitional City

Shenzhen's push to progress politically is tempered by the challenges born of earlier success and present within and beyond its borders. Governing a city of almost 14 million people in a built-up area of 720 square kilometers (in contrast to only three in 1979) is daunting. It is a planned city, but still suffers from a lack of overall planning and a failure to anticipate the problems of its scale. Furthermore, municipal policymakers often favor speed over coordinated growth. Until recently, government bureaucracy has made it both difficult and expensive for migrant workers to obtain the temporary residential permit. In response to the higher proportion of migrants involved in crimes, Shenzhen's police have been heavy handed in cracking down by rounding up people perceived as migrants with legitimate temporary residence. The speculative real-estate market that has developed in response to the rapid economic growth has pushed up housing prices in central Shenzhen to an almost comparative level with Hong Kong's New Territory across the border. This has forced migrant workers to the city's periphery to rent or to build the equivalent of slum housing on ecologically fragile hills. The *Asia Times* reports that upwards of four hundred thousand migrant laborers were involved in building and living in makeshift housing, mostly on Shenzhen's outskirts, against government regulation.[18] Shenzhen was built with wider roads to accommodate cars but without attention to the walking needs of pedestrians and access to public buildings and spaces. Pedestrians can only cross the streets at pedestrian overpasses, which are useless for wheelchairs, making them discriminatory towards the disabled.[19]

The insufficient municipal social services for residents and migrants have stimulated bottom-up political aspirations, especially with the growing middle class in Shenzhen. When Mr. Jiang Shen ran for political office to challenge an election law he saw as corrupt, he used his cell phone to find supporters for his candidacy and distributed campaign cards to neighbors. He met opposition from state officials, however, who were disposed against him because of a lawsuit he had brought against a city official (French 2006). His efforts demonstrate that urbanism creates new civic individualized mind-sets. Within the system, he pursued avenues to enact change in a manner that made it clear that this was an individual acting. By utilizing personal technology and grassroots tactics, his efforts are replicable by other individuals. While he did not win, his replicable efforts show that urban identity calls for greater civic participation.

Global forces, however, have hampered deep governance reform. The 2008 economic crisis eroded the government's willingness to enforce China's new labor law, which provided more effective protection for migrant laborers' interests and rights effective January 1, 2008, as well as reform as a whole. By October of 2008, 15,661 enterprises in Guangdong province shut their doors, and over half of those—about 8,500—ceased doing business in that month alone. The central government's new permission for local authorities to freeze minimum-wage levels and to reduce or suspend employers' social-insurance contributions compromised the willingness and ability of the Shenzhen government to protect migrant workers' interests.

As Shenzhen shifts from an "instant city" to a more global secondary city, its governance becomes more complex due to the interaction of internal and external dynamics. Low taxes, massive Hong Kong investment, cheap prices, and government subsidies were huge catalysts for the fast growth of labor-intensive and export-dependent industries. However, upgrading to higher value-added manufacturing and services necessitates more strategic policy coordination and sophisticated handling of state-market relations. After much delay, the government has taken a bold new step by issuing the permanent residential card to millions of migrants, but without specifying and guaranteeing that they will have the same rights as the small percentage of the population with *hukou*. The global economic crisis has raised the stakes for more effective governance response, yet has also strengthened a government that attempts to shrink itself through ongoing reform.

THE PROSPECT FOR, AND THE LESSONS LEARNED, FROM SHENZHEN

Shenzhen's first-mover role will continue to be present in the DNA of its maturation into a second-tier global city. Whereas sudden growth may have turned Shenzhen into an urban anomaly—a city without historical identity—its short history as a SEZ is undoubtedly its dependent path. The Shenzhen SEZ was the *first mover* of economic reform and open policy during the early stage of China's transition to a hybrid socialist and capitalist system. Shenzhen set the tone and stage for the gradual and cumulative process that unfolded over the next three decades while China rose as a global player. Shenzhen became *the* experimental ground for FDI, joint ventures, land tendering, contractual employment, the blurring of urban and rural distinctions through migration, and more. When the successful experiments became transplantable and replicable in other areas, Shenzhen began to lose its special status. Shenzhen was the first domino that Deng Xiaoping lined up in his row. It was the first domino that triggered the row's fall, but also lost some of its original distinction once its counterparts fell upon each other. Even so, it retains its unique status in that it was the beginning that dictated the fate of the whole lot—for better or for worse—of dominoes that followed.

Beyond its first-mover status and modeling role as a SEZ, Shenzhen has grown into a huge industrial city confronting new challenges that threaten its continued prosperity. World-renowned architect Rem Koolhaas remarked, "The absence, on the one hand, of plausible, universal doctrines and the presence, on the other, of an unprecedented intensity of production, have created a unique, wrenching condition: the urban seems to be least understood at the moment of its apotheosis" (cited in Ratti 2003: 210). Shenzhen lacks a deeper and more diverse economic base given its industrial origination, even though it has weathered the recent global recession quite well. Hong Kong has simultaneously fostered rapid growth through the convenient cross-boundary transfer of massive capital for export-oriented production and has also reinforced Shenzhen's shallow and single-dimensional economic strength. As Hong Kong's own disadvantages—high salaries, expensive land and the overwhelming dominance of the service and real-estate sectors—persist, it may be less capable of helping Shenzhen achieve successful industrial upgrading through the current global economic downturn. Simply put, Hong Kong's advantages have been eroded by the faster overall growth and upgrading of Guangdong and neighboring cities, especially Shenzhen since 2001 (Shen, 2008). Speaking more of the Guangdong–Hong Kong competition, the Bauhinia Foundation Research Centre in Hong Kong predicts that within ten years Guangdong will exceed the per capita GDP, in aggregate terms, of Hong Kong.[20]

Nevertheless, in meeting the new challenges of rising cost and industrial upgrading, Shenzhen has (re)turned to its old neighbor—the true global city of Hong Kong—as a key component of Shenzhen's 2030 Urban Development Strategy, which envisions the city becoming a global urban center for sustainable development (Shen 2008). This is the first time Shenzhen has taken economic integration with Hong Kong into consideration for long-term development. Yet, this integration is appropriate given the impressive daily traffic, on average, across the two land borders, which numbered 505,000 in 2010.[21] The draft blueprint, drawn up in late November 2007 and sent to the State Council (China's cabinet) for approval, called for Shenzhen and Hong Kong to convert themselves into a twin-financial, trade, and shipping hub. One of the priorities was to build a common capital market with Shenzhen's financial institutions going international through Hong Kong, while more financial institutions in the former British territory were to set up branches in Shenzhen to expand business in the mainland. The Shenzhen blueprint focused on six major areas: improving cooperation on financial systems; building a Shen(zhen)–(Hong) Kong innovation rim; improving cross-border transportation; and enhancing cooperation with Hong Kong in the high-technology and high-end service industries. To build the Shen-Kong metropolis, the Shenzhen government reiterated that it was willing to become "the backyard" of Hong Kong as the region's international financial hub. The draft regulation was aimed at boosting Shenzhen's financial sector into "a strategic pillar industry" of the city after high-technology manufacturing (Chung 2007).

As rosy a scenario as the blueprint depicted, there are heavy barriers to be overcome before execution. For example, whereas the Chinese currency is in some circulation in Hong Kong, it is not fully convertible, which greatly restricts capital flow between Shenzhen and Hong Kong. Legal systems in the two places are also radically different, so to build a common capital market, a common legal basis must be established. (See Liu and Chen on similar and different barriers to linking Tianjin and its dominant neighbor of Beijing in the next chapter.) The Bauhinia Foundation Research Centre in Hong Kong posited in its report of August 2007 that cross-border growth and integration could be facilitated with a "Hong Kong-Shenzhen one-hour metropolitan life cycle" through efficacious usage of rail and roads to connect the two metropolises at multiple spots along their borders.[22]

If Shenzhen is to successfully transition from an "instant city" to a sustainable secondary global city that it aspires to be, it needs to develop a functioning protocol for dealing with millions of migrant laborers. Although most of them have lived in Shenzhen for a long time with or without permanent residence, many have yet to feel at home in and develop an attachment to Shenzhen. Even if the recent global economic crisis provided a temporary relief to this alienation as many migrants were forced to leave Shenzhen to return to their places of origin, this lack of identity with Shenzhen is a roadblock to its transition to a more sophisticated city. Moreover, some of the more skilled and educated temporary workers (Shenzhen boasts the highest percentage of PhDs of all Chinese cities) will not be fully committed to working in the city. Having begun to issue the permanent residential card to the millions of migrant workers in 2008, the Shenzhen government has taken a great leap forward in governing the transitional city as a more equitable and sustainable living place. In this crucial sense, Shenzhen is an unusual "instant city" because its "coming of age" highlights both the dynamics and tensions of transitioning to a secondary global city, as it is unlikely to ever outshine Shanghai and Beijing. It even remains to be seen how Shenzhen will fare in comparison to Tianjin, another second-tier Chinese city aiming to play on the global stage. That is the subject of the next chapter.

ACKNOWLEDGEMENT

The first draft of this paper was presented at the conference on "Rethinking Cities and Communities: Urban Transition before and During the Era of Globalization," sponsored by the Center for Urban and Global Studies at Trinity College, November 14–15, 2008. An abridged revision of the conference paper was published as Xiangming Chen and Tomás de'Medici (2010), "The 'Instant City' Coming of Age: The Production of Spaces in China's Shenzhen Special Economic Zone." *Urban Geography* 31 (8): 1141–1147.

NOTES

1. On scale, Dubai's population grew from 25,000 in 1948 to about one million around 2000 (Kotkin 2005), which was dwarfed by Shenzhen's explosion from around 30,000 in 1979 to over 10 million in 2009. On function, a small and remote port in the sprawling British "informal empire" until the 1970s, Dubai has since been marked by a greater variety of economic activities such as smuggling, transshipment, free-trade zones, and, more recently, financial services (Kanna 2011).
2. T. K. Maloy (2003), "Shenzhen: The Instant City," online news article originally published on United Press International.com, October 1, accessed on April 20, 2009.
3. "Records of comrade Deng Xiaoping's Shenzhen Tour," *The People's Daily* online accessed on http://english.peopledaily.com.cn/200201/18/eng20020118_88932.shtml, accessed on April 10, 2009.
4. This calculation excluded the 6 million temporary migrant population, which helped build Shenzhen's industrial machine (Wang 2008).
5. Same as note 2 above.
6. "Shenzhen's Role Adjusted to Spearhead Reform," *Xinhua News Online*, August 26, 2010.
7. *Asia Times* online at www.atimes.com, accessed on November 22, 2007.
8. Alan Wheatley, "China's Growth Slowly Starts to Shift Inland." Accessed on http://www.reuters.com/article/newsOne/idUSTRE53C0LN20090413?sp=true, April 13, 2009.
9. See note 5 above.
10. The reporter who raised the question published the controversial article on November 17, 2002, and then left Shenzhen for work in Hong Kong. We accessed the article on www.cul.book.sina.cn/view/shzh/ on January 15, 2009.
11. "Letting sparrows go and inducing phoenixes: Shenzhen's turnaround," *The People's Daily*, December 3, 2007: 6.
12. "Miracle Shenzhen," *The People's Daily*, November 26, 2008: 2.
13. "The Spirit of Shenzhen," *The Wall Street Journal Asia*, January 9–11, 2009: W12.
14. This card can be issued to all Shenzhen temporary residents over 16 years of age who are employed, have investment, own properties, are overseas returnees, or possess special creative talents. The government also introduced a "temporary residential card" for those who do not have jobs, investment and private property in Shenzhen.
15. "The flag for reform and opening," *The People's Daily*, October 1, 2008: 2.
16. "Shenzhen: Mayoral Election in Three Years." *First Fiscal and Economics Daily*, May 26, 2008: 5.
17. "The planned elimination of the district government is Shenzhen's pioneering role in governance reform," *First Fiscal and Economics Daily*, November 24, 2008: 4.
18. See note 6 above.
19. Sean Marshall, "Shenzhen: China's Instant City," *Spacing Toronto*, October 13, 2008, accessed on April 20, 2009.
20. *The China Daily* online at http://www.chinadaily.com.cn/hkedition/2011-08/31/content_13224634.htm, accessed on September 1, 2011.
21. "Hong Kong-Shenzhen Integration Plan Progressing." Accessed from http://nextbigfuture.com/2011/09/hong-kong-shenzhen-integration-plan.html, September 9, 2011.
22. Ibid.

REFERENCES

Braudel, Fernand (1992). *Civilization and Capitalism 15th–18th Century: The Perspective of the World*. New York: Harper & Row Publishers.

Brookings Institution and London School of Economics (2010). *Global Metro Report: The Path to Economic Recovery*. A report by the Metropolitan Policy Program of Brookings and The Cities Center of LSE on the 150 global metropolitan economies in the wake of the great recession. December.

Chen, Xiangming (1987). "Magic and Myth of Migration: A Case Study of a Special Economic Zone in China." *Asia-Pacific Population Journal* 2: 57–77.

Chen, Xiangming (1994). "The Changing Roles of Free Economic Zones in Development: A Comparative Analysis of Capitalist and Socialist Cases in East Asia." *Studies in Comparative International Development* 29: 3–25.

Chen, Xiangming (1995). "The Evolution of Free Economic Zones and the Recent Development of Cross-National Growth Zones." *International Journal of Urban and Regional Research* 19: 593–621.

Chen, Xiangming (2007). "A Tale of Two Regions in China: Rapid Economic Development and Slow Industrial Upgrading in the Pearl River and the Yangtze River Deltas." *International Journal of Comparative Sociology* 48: 167–201.

Chen, Xiangming, ed. (2009). *Shanghai Rising: State Power and Local Transformations in a Global Megacity*. Minneapolis: University of Minnesota Press.

Chen, Xiangming, and Tomás de'Medici (2010). "The 'Instant City' Coming of Age: The Production of Spaces in China's Shenzhen Special Economic Zone." *Urban Geography* 31 (8): 1141–1147.

Chung, Olivia (2007). "Shenzhen Nuzzles Closer to Hong Kong: Two Parts." *Asia Times* online at www.atimes.com, accessed December 5 and 7.

Farole, Thomas (2011). "Special Economic Zones: What Have We Learned?" *Economic Premise* 64 (September): 1–5.

French, Howard W. (2006). "Shenzhen's Citizens Are Defending Their Rights." *International Herald Tribune*, December 17.

Hu, Mou (2007). "The Acceleration of Shenzhen's Industrial Upgrading." *People's Daily*, December 3: 6.

Kanna, Ahmed (2011). *Dubai, the City as Corporation*. Minneapolis: University of Minnesota Press.

Kotkin, Joel (2005), *The City: A Global History*. London: Weidenfeld & Nicolson.

Li, Luoli (2008). "The Definition and Direction of Shenzhen's Reform and Opening in a New Era" (in Chinese). *China Opening Herald* 1: 5–8.

Liang, Zai (1999). "Foreign Investment, Economic Growth, and Temporary Migration: The Case of Shenzhen Special Economic Zone, China." *Development and Society* 28: 115–137.

Nishitateno, Sonoko (1983). "China's Special Economic Zones: Experimental Units for Economic Reform." *International and Comparative Law Quarterly* 32: 175–185.

Ouroussoff, Nicolai (2008). "The New, New City." *The New York Times Magazine* (June 8): 70–75.

Ratti, Carlo (2003). "Instant City (Shenzhen, China)." Unpublished paper by the SENSEable City Laboratory, Massachusetts Institute of Technology.

Roberts, Dexter (2009). "As Factories Fail, so Does Business Law." *BusinessWeek* (April 13): 46–48.

Shen, Jianfa (2005). "Counting Urban Population in Chinese Censuses 1953–2000: Changing Definitions, Problems and Solutions." *Population, Space and Place* 11: 381–400.

Shen, Jianfa (2008). "Hong Kong under Chinese Sovereignty: Economic Relations with Mainland China, 1978–2007." *Eurasian Geography and Economics* 49: 326–340.

Shenzhen Statistical Bureau (1985). *Shenzhen Statistical Yearbook 1985*. Shenzhen: Shenzhen Statistics Press

Shenzhen Statistical Bureau (2010). *Shenzhen Statistical Yearbook 2008*. Shenzhen: accessed on www.sztj.com, September 1, 2009.

Sit, Victor F. S. (1988). "China's Export-Oriented Open Areas: The Export Processing Zone Concept." *Asian Survey* 28: 661–675.

Wang, Chao (2008). "The Story of Spring: Shenzhen's City-building History." Excerpted from *The China Youth Daily* online at www.china.youth.com, April 22.

Xiao, Jin (2003). "Redefining Adult Education in an Emerging Economy: The Example of Shenzhen, China." *International Review of Education* 49: 487–508.

Yeung, Yue-man, Joanna Lee, and Gordon Kee (2009). "China's Special Economic Zones: Three Decades of Changing Roles and Achievements." *Eurasia Geography and Economics* 50: 222–240.

You, Jin (2008). "No More Outsiders in Shenzhen" (in Chinese). *Government Law* 19: 16–17; downloaded from China Academic Journal Electronic Publishing House at www.cnki.net, September 1, 2009.

Zhang, Guodong (2007). "Shenzhen: A City of Immigrants" (in Chinese). *Xiaokang* 12: 36–38; downloaded from China Academic Journal Electronic Publishing House at www.cnki.net, August 1, 2010.

7 The Third Coming of China's Special Economic Zones

The Rise and Regional Dimensions of Tianjin Binhai New Area

Chang Liu and Xiangming Chen

INTRODUCTION

Since becoming the newest national economic zone and growth engine of China in 2005, Tianjin Binhai New Area (hereafter TBNA) has drawn attention across China but remains relatively unknown to the outside world. Located on the coastline of Tianjin Municipality (see Map 7.1), the largest industrial city in northern China, TBNA has been one of the fastest growing local economies in China with gross domestic product (GDP) averaging about 20 percent since 2006. The Chinese media has dubbed it as "China's third growth engine." It follows the special economic zone (SEZ), and now megacity, of Shenzhen bordering Hong Kong that took off in the 1980s (see the preceding chapter) and the Pudong new district of Shanghai that has flourished since the early 1990s (Chen 2009). We see TBNA as the third coming of China's SEZs, with important retrospective connections to Shenzhen and Pudong and prospective implications for major cities in China and beyond.

TBNA is not entirely new. Its core area, Tianjin Economic Development Area (TEDA), was founded in 1984. After Shenzhen was opened to foreign investment around 1980 and enjoyed development success almost overnight (Chen and de'Medici 2010, this volume), the central government of China set up more SEZs along the entire coast in 1984, and TEDA was among them. For a long time, however, TEDA did not make the expected strong contribution to the economy of Tianjin, whose annual GDP growth remained at about 7 percent, well below the national average of nearly 10 percent through the early 1990s.

Faster development of TEDA did not occur until 1992, when global forces began to exert a stronger local impact. Regarding the growing investment by multinational companies in TEDA, Motorola sunk $120 million and built a huge mobile phone factory, which was the largest foreign investment project in China at that time. As Samsung, Panasonic, and other large electronic companies followed suit, TEDA became number one in attracting foreign investment among China's economic development zone for nine years in a row from 1993 (Liu 2008). This infusion of global capital contributed to an annual GDP growth of 20 percent in TBNA during the 1990s.

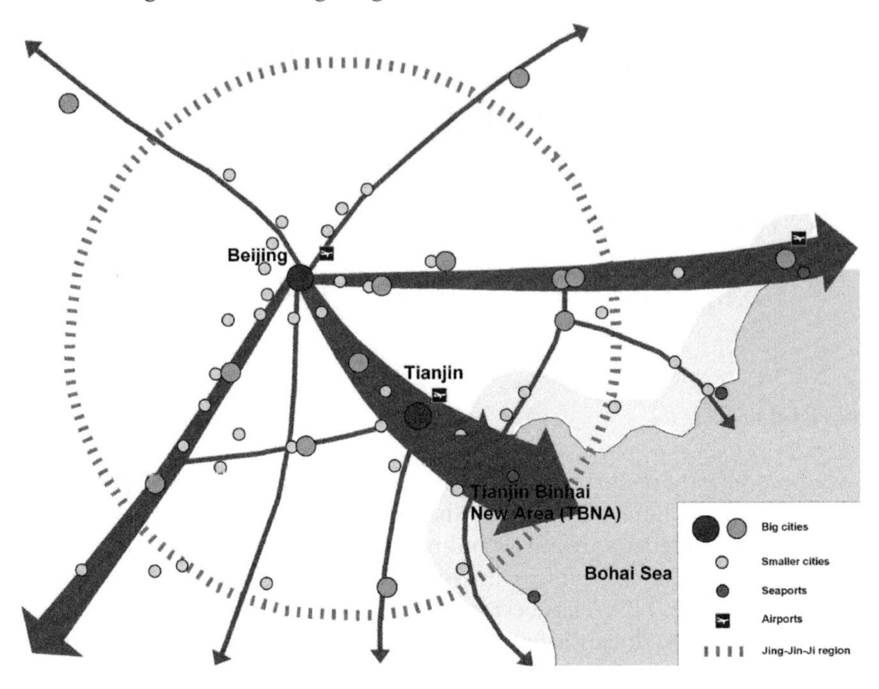

Map 7.1 Tianjin Binhai New Area (TBNA) in the Beijing-Tianjin-Hebei (Jing-Jin-Ji) region.
Source: Redrawn by Gordon Kee from Wu (2005).

In 1994, the municipal government of Tianjin proposed to set up the Binhai New Area. With TEDA as its core, TBNA would cover 2,270 square kilometers with seven districts of separate functional orientations. It remained an administrative collection of the separate districts without a powerful or enforcing governing unit until 2006, when the State Council of China approved TBNA to be an experimental zone for comprehensive reform and declared the inclusion of TBNA into a grand national development strategy via *On Problems Concerning the Promotion of Development and Opening of Tianjin Binhai New Area* (Document No. 20). This document stated the goals of TBNA as: "1) being based in the Beijing-Tianjin-Hebei (province) region; 2) serving the Bohai Rim on the Yellow Sea; 3) spreading the development benefits to northern China; and 4) connecting to Northeast Asia and become the gateway to northern China."[1]

These ambitious goals were feasible through the central government's top-down push given its unchallenged authority and abundant resources as proven by Shenzhen in the 1980s and Pudong in the 1990s. This chapter examines the rise and regional dimensions of TBNA. First, we place the case of TBNA in relevant theoretical context. Then we analyze the empirical information regarding TBNA in a broader framework involving such reference or comparable cases as Shenzhen and Pudong, Shanghai. Finally, we draw limited conclusions and policy implications based on the evidence.

CITY-REGIONAL DEVELOPMENT IN LOCAL CONTEXT

Urban economists hold that cities are results of agglomeration effects and increasing return to scale. This basic condition attracts firms because of the positive externalities of infrastructure sharing, labor matching, and knowledge spillover within and between spatial clusters. Among these externalities, knowledge and technology would be a major source of economy-wide increasing return and thus the source of long-run economic growth under resource constraints (Romer 1990, cited in Storper 2009). (See Forrant, this volume, on the skill- and technology-driven development of the machine-tool industry in Springfield, Massachusetts, in the late nineteenth century.) Romer's argument is often linked with the earlier contributions by Marshall (1920) and Arrow (1962), who focused on the local implications of technology spillovers, which became known as the Marshall-Arrow-Romer Externality Theory for economic growth. In this process, innovation is first created by firms in some regions of technology advantages, and for a certain amount of time, monopoly rent is enjoyed from the invention as long as the barrier remains. However, as the innovation is more widely used over time, substitute technology may be developed and imitation begins to take place.

If this is the only process, as Storper (2009) suggests, we should find a homogeneous world that develops at about the same pace and in the same direction. However, in the real world, the global division of labor and fragmentation of production requires further theoretical discussion. Specific innovation only happens in certain places, and some places, the so-called learning regions, can imitate better than others. This presses us to probe key differences among regions. To mainstream economists, actors and decision-making processes are the same everywhere, but preferences, endowments, and factor costs are different; whereas sociologists and anthropologists would see an economic actor's decision-making process as embedded in the tradition and institutions of localities. Storper (2009) defined such different traditions and institutions across places as *local contexts*. This theorizing of local context can be linked to other empirical studies to help us understand regional differences that may be hidden in the regional "black box" (Florida, Mellander, and Stolarick 2007).

Going back to the case of TBNA, the idea of local context can help explain its advantages and disadvantages in intra- and interregional and global competition. In local institutional and cultural terms, TBNA is relatively new and thus has little tradition and traction. This means that local firms are not fixed into one particular way of thinking and do not have much bias in decision making about how to innovate. Technologically speaking, TBNA is not at the frontier of global or even regional innovation. Its success depends largely on how it receives the diffusion of technology advances, and its imitation of such innovations. That requires two conditions: one is the presence of multilocation companies with new technology. Whereas the central and local governments' policies have attracted many multinational firms such as

Airbus and Motorola, the most advanced technology may not be brought to and so easily digested by the TBNA firms and factories, which may have to climb a steep learning curve toward innovative practices.

The second condition involves sufficient local human capital and the presence of research institutes. Human capital is needed for both receiving global diffusion of knowledge and imitating such technology advances, which can be cultivated by strong and sophisticated research universities and institutes. This condition becomes more significant for TBNA to further its "original innovations," as promoted by the central government. However, TBNA does not host many prominent research institutes. Although several of China's highly ranked universities are in Tianjin and certainly in nearby Beijing, which is China's best educational and research and development (R&D) center, TBNA itself has not developed long enough to have this human capital base (An and Chen 2009). Instead, TBNA needs to adopt a more flexible approach to attract the wider human capital flows into and spanning the broader region.

FROM GLOBAL CITY TO GLOBAL CITY-REGION

The global city perspective, advanced by sociologist Saskia Sassen (1991), looks at cities as interrelated nodes embedded in a global economic network and urban hierarchy. At the top are New York and London, with their commanding power on the global economy, whereas lower down are globalizing cities with different levels and strength of global connectivity and functional influence. This view turns the assumption that the city is a localized and bounded territory on its head by conceptualizing it as partially denationalized or detached with significant autonomy from the restrictions of national government.

While not a global city like New York or London, TBNA is fast becoming a new key node in the global economy. Hosting one of the busiest container ports in China and a busy international airport, TBNA is capable of serving as the gateway to a much larger city-region around Tianjin. Its favorable location in northeast Asia attracts a lot of Korean and Japanese capital (OECD 2009). From the global city perspective the question of whether and how growing global connections, coupled with the special zone status like Shenzhen or Pudong, can make TBNA somewhat autonomous from its national political and economic anchors and with what consequences arises.

Unaligned with conventional national or local political jurisdictions, a global city can spill and extend economic activities and information impulses into its hinterlands and beyond (Ren, Chen, and Läpple 2009). In global cities, service industries are densely concentrated in the central business district (CBD), which helps push out other industries and some residents to suburbs, even to some nearby cities with lower-cost labor or land. This process stimulates the emergence of the global city-region, which may encompass multiple cities that are functionally specialized and linked through their comparative advantages and transport routes. The global city-region, as Vogel (2010) suggests, can serve as a more precise description for Sassen's global city. Moving from the global city to the global city-region casts the spotlight on the distinctive *regional contexts* (relative to the local contexts discussed earlier) of large globalized or globalizing urban entities. As Brito and Correia (2010) suggest, the geographic proximity of many firms does not automatically equal an economy of scale, but some mechanism must be created to foster a positive relationship among market actors to make agglomeration beneficial. This logic, ironically, may also cross over to the intergovernment relationship at the regional scale. Vogel (2010) advocates the use of "regional governance" to hold many regional administrative units together where fierce competition can otherwise lead to fragmentation. While seemingly collaborating in a cross-border regional context, the Detroit-Windsor metro area has been undergoing more de-globalization rather than globalization at the local scale (see Ryan, this volume).

The Beijing (Jing)–Tianjin (Jin)–Hebei (known as Ji in Chinese official use) region (hereafter Jing-Jin-Ji), where TBNA is located, is among the world's major city-regions in terms of territorial, demographic, and economic scales. In research using light emission-related economic activities as a major indicator, the Jing-Jin-Ji region ranks the thirty-fourth worldwide, ahead of the Greater Berlin region and Singapore (Florida, Gulden, and Mellander 2007). The Jing-Jin-Ji region is sometimes referred to as Greater Beijing by Chinese scholars (e.g., Wu 2005), because Beijing constitutes the central hub for the region. However, the idea of Greater Beijing has not caught on with TBNA, which instead identifies itself as a major and distinct regional center. This is the result of both the historical competition between Tianjin and Beijing and the explosive economic development of TBNA. Besides this competition, the less developed secondary cities of Hebei province, whose economy largely depends on agriculture, raw material exports, and low-end manufacturing, may become an obstacle for intraregional cooperation. Whereas there exists severe uneven development between Hebei province and the two dominant municipalities of Beijing and Tianjin, that all three are provincial authorities creates both unequal and yet comparable relations that challenge intraregional cooperation. This regional context is fitting for a multitiered and multidimensional analysis of TBNA.

TBNA IN MULTIDIMENSIONAL AND COMPARATIVE PERSPECTIVES

Explosive Growth and Contributing Factors

One can hardly dispute TBNA as being one of the fastest-growing areas in China over the last decade or so. After averaging 20 percent in annual growth, TBNA's GDP for 2009, when the world only began to recover from the crisis and China's GDP growth fell to only 8.7 percent, rose 23.5 percent over the previous year (see Figure 7.1). Besides this miraculous cumulative growth, it was shocking that TBNA's total GDP surpassed that of Pudong, the growth leader of China's economic zones since the early 1990s, by over $3 billion in 2008. As Figure 7.1 also shows, the dynamic growth of TBNA was fueled by a large influx of foreign investment, even through the crisis year of 2009. In 2006 and 2007, foreign-funded enterprises accounted for 20 percent and 16 percent of TBNA's total fixed asset investment, respectively. A chunk of this heavy foreign investment was accounted for by the establishment of a highly capitalized Airbus assembly facility, one of the four in the world and the only one in a developing country.

To explain TBNA's rapid growth, Chinese scholars have studied a variety of its comparative and competitive advantages such as *location*, on the coast with port facilities (Chang 2007; Lei 2007; Li 2008); *transportation*, well connected to the center of Tianjin and beyond (Lei 2007; Wu 2007); *natural resources*, especially land resource, with 2,200 square kilometers of land for development (Lei 2007; Li 2008; Xiao 2006); *human capital*, an easy access to university graduates in Tianjin (Li 2007; Xiao 2006); and *industrial foundation*, linked to Tianjin's strong manufacturing tradition (Chang 2007; Lei 2007; Xiao 2006). Regarding location, TBNA is also China's gateway to northeast Asia with its geographic proximity to Japan and Korea, which attracts incoming and outgoing investment with the two developed economies. Secondly, geographic proximity to the capital city of Beijing—China's educational, technological, and second economic center (behind Shanghai)—fosters cooperative opportunities. With its location advantages enhanced by

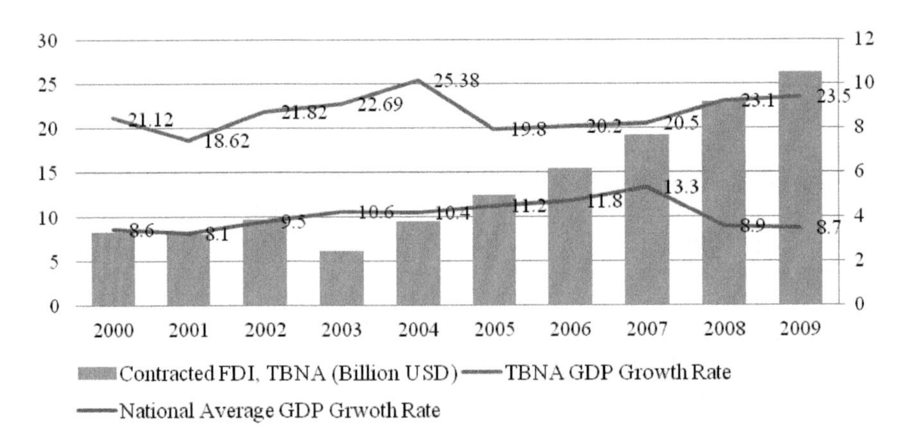

Figure 7.1 GDP growth rate, national and TBNA, with contracted foreign direct investment in TBNA, 2000–2009.
Sources: Statistic Yearbook of China 2008, 2010, Statistical Yearbook of Tianjin 2008, 2009, 2010. Development Report of TBNA, 2007.

favorable transport infrastructure, TBNA benefits further from the rapid growth of the Tianjin Port, which in 2008 passed Guangzhou and became China's third busiest harbor for cargo shipping, behind only Shanghai and Shenzhen, rising to the world's fifth busiest port.[2]

Building off Tianjin's manufacturing base, TBNA has developed strong comparative advantages in a long list of industries, including electronics, information, petroleum, chemical, automobile, steel, bio-tech, new energy, new materials and green industries (Zhu and Sun 2009). This industrial base, with the positive externalities created by the diversified and intertwined industries, facilitates current and future economic growth through continued agglomeration and diversification.

Besides these "natural" entrenched advantages, TBNA has enjoyed recent advantages "manufactured" by the state, the most significant of which was aggressive promotion by the national government. In October 2005, the central government placed TBNA in the same league as Shenzhen and Pudong (Zhu and Sun 2009). In the following year, the State Council announced TBNA as an "experimental zone for comprehensive reforms." Since then, state media have kept TBNA in the limelight and touted it as the "third growth engine" on a number of important occasions. Such top-level promotion has helped draw major international investors, highlighted by the establishment of the Airbus assembly line in 2007, which would not have materialized in TBNA without the central government's targeted promotion and pull.

Moreover, the government has injected substantial investment into local and regional transport infrastructure. The fast train that connects Beijing, downtown Tianjin, and TBNA absorbed an investment of nearly $2 billion. The train allows people to travel from TBNA to downtown Tianjin in fifteen minutes or to Beijing in forty-five minutes. In 2009, the TBNA administration began the so-called Ten Campaigns in order to improve the investment environment. The official Web site of TBNA claims that new funding would reach a stunning amount of $200 billion.[3] By 2009, in the midst of the global financial crisis, the campaigns resulted in an investment of more than $15 billion, materialized as new highways, a power plant, entertainment facilities, and other physical infrastructure for business and industries that would exert a stronger impact on TBNA's local economic structure.

Economic Structure in Comparative Perspective

As the manufacturing and infrastructure-driven growth of TBNA has accelerated, the central and local governments have also ratcheted up their advocacy for developing service industries in TBNA. This ambitious dual goal, set and advocated by the state, is up against the constraints of path-dependent development, more visibly in comparison to Pudong and Shenzhen over the more recent period. As Figure 7.2 reveals, the manufacturing sector not only dominated the TBNA economy but also gained a little ground relative to services after 2006. In comparison, Pudong, which boasted a larger manufacturing sector in 1993, right after its own national launch as a new economic zone, grew its service sector to the point where it proportionally surpassed manufacturing in 2007. As Bai and Li (2003) contend, Pudong had entered a "post-industrial stage," and became a service provider for Shanghai's broad hinterland of the Yangtze River Delta (Chen 2009). Regarding the other close reference case, Shenzhen's industrialization took off in the 1980s and accelerated through the 1990s. Although still a major manufacturing center, Shenzhen has achieved and maintained an evenly balanced local economy of manufacturing and services since the early 2000s (Chen and de'Medici 2010).

The evidence points to a longer evolution and adjustment of Pudong's and Shenzhen's economic sectors given their earlier start. TBNA, however, remains locked into its manufacturing groove, reinforced by the fact that the bulk of the fixed-asset investment in 2009 was channeled into manufacturing. At the end of 2009, TBNA managed to get the central government's approval of their plan to experiment with "financial innovations."[4] Such innovations would include private equities, venture investments, offshore banking, and other high-end services that were severely restricted by the national government. TBNA's service sector is dominated not by financial services but by transportation and logistical services. The fate of the financial-services sector remains uncertain due to its minimal role in the region's tertiary sector. Overall, TBNA differs from both Pudong and Shenzhen in the relative strength of manufacturing versus services due to their varied stages of development and changing external environments, which includes the shadow influence of nearby Beijing.

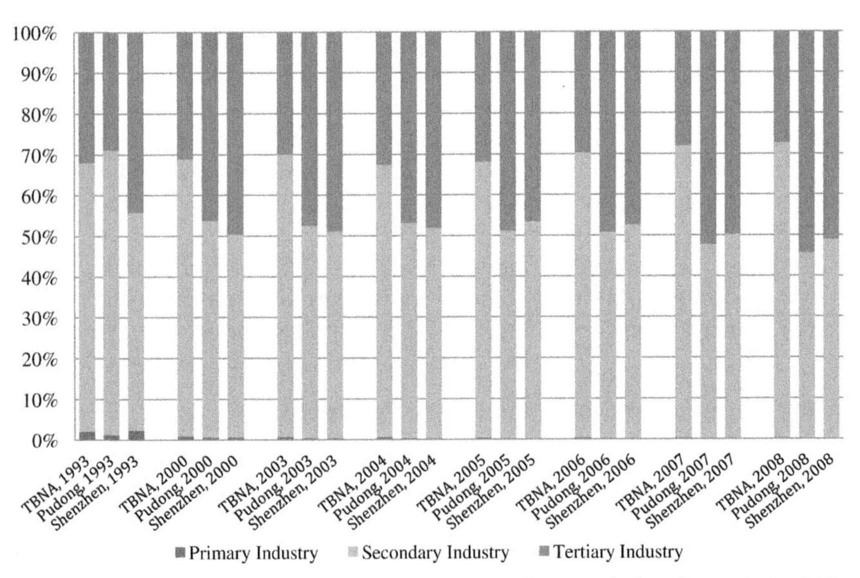

Figure 7.2 The industrial composition of TBNA, Pudong, and Shenzhen, 1993–2008.
Sources: Statistical Yearbook of Tianjin 2004-2009. Statistical Yearbook of Shanghai 2000-2009. Statistic Yearbook of Shenzhen, 2009.

The Beijing Effect: A Blessing or a Curse?

For TBNA, the close proximity to Beijing is a double-edged sword. Complementing its status as the political center of China, Beijing is above or comparable to Shanghai as an educational and innovation hub. In 2008, Beijing produced more than 6,478 patents, first among all Chinese cities, whereas Shanghai occupied the second place with 4,258 patents. With the largest concentration of top universities and institutes, Beijing leads all Chinese cities in the abundant supply of highly educated and specialized labor force. In addition, Beijing and Shanghai host a comparable amount of multinational corporations' headquarters and international banks, but the former's capital exceeds the latter's, with more large state-owned enterprises headquarters.

Given Beijing's dominance and ambition, an ideal Beijing-Tianjin division of labor would lead TBNA to position itself to receive the manufacturing industries that Beijing is willing to let go and to develop a "designed in Beijing, made in Binhai" industrial nexus (Wu 2007). Hewlett-Packard's recent actions illustrate this relationship. In 2010, the firm decided to open a $5 million cloud computing and data center in TBNA. This new center will most likely become an anchor for a whole chain of suppliers and manufactures to follow into TBNA. With headquarters in Beijing, HP can combine the cheaper land and preferential policy of TBNA with the better R&D facilities and high-quality human capital of Beijing, conveniently linked by the forty-five-minute fast train service. This "Beijing effect," however, may cut both ways. It means that TBNA will have to compete with Beijing for both financial and human capital (Zhu and Sun 2009), albeit from a less advantaged position.

TBNA's delicate position in the Beijing-Tianjin nexus is revealed further by the two cities' divergent development paths (see Figure 7.3). Despite their

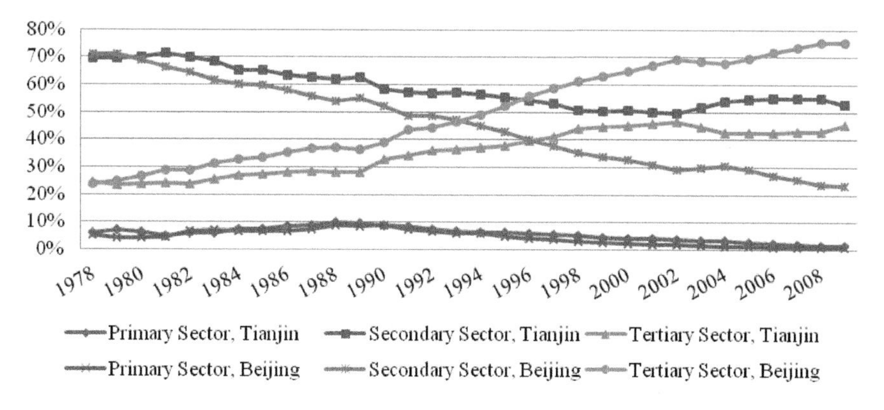

Figure 7.3 Trends in the industrial composition of Beijing and Tianjin, 1978–2009.
Sources: Statistic Yearbook of Beijing 2010, Statistical Yearbook of Tianjin 2010.

Table 7.1 Gross Domestic Product (GDP) for the Industrial Sectors of Beijing and Tianjin with Service-Sector Breakdowns, 2003–2009 (in Billion U.S. Dollars*)

	Year	2003	2004	2005	2006	2007	2008	2009
Beijing	Gross Domestic Product	77.6	93.5	108.1	125.9	152.7	172.3	188.4
	Agricultural Sector	1.3	1.3	1.4	1.4	1.6	1.8	1.8
	Manufacturing Sector	23.1	28.7	31.4	34.0	38.9	40.7	44.3
	Service Sector	53.3	63.4	75.3	90.5	112.2	129.9	142.3
	Transportation, Storgae, Post and Telecommunications	4.8	5.5	6.2	7.1	7.7	7.7	8.6
	Information Transmission, Computer Services and Software	5.9	7.0	9.1	10.8	13.5	15.5	16.5
	Wholesale and Retail Trade	8.0	9.1	10.9	13.5	17.0	22.1	23.6
	Finance	9.9	11.1	13.0	15.2	20.2	23.6	24.9
	Real Estate	5.3	6.8	7.7	10.2	12.7	13.1	16.5
	Tenancy and Commercial Services	3.6	4.3	5.6	6.9	9.7	11.9	12.6
	Scientific Research, Technical Services and Geological Prospecting	3.8	4.3	5.6	6.9	9.7	11.9	12.6
	Other Industries	12.1	15.4	17.3	20.0	22.6	25.0	27.0
	Finance as % of GDP	**12.7%**	**11.8%**	**12.1%**	**12.1%**	**13.2%**	**13.7%**	**13.2%**
	Finance as % Service GDP	**18.5%**	**17.4%**	**17.3%**	**16.8%**	**18.0%**	**18.1%**	**17.5%**
Tianjin	Gross Domestic Product	38.0	48.2	57.3	67.6	81.4	104.2	116.6
	Agricultural Sector	1.4	1.6	1.7	1.8	1.7	1.9	2.0
	Manufacturing Sector	19.3	26.1	31.8	38.6	44.9	57.5	61.8
	Service Sector	17.2	20.5	23.8	27.2	34.9	44.8	52.8
	Transportation, Storage, Post and Telecommunications	3.0	3.5	3.5	3.9	5.2	6.8	7.3
	Information Transmission, Computer Services and Software	1.1	1.0	1.2	1.3	1.4	1.7	2.1
	Wholesale and Retail Trade	2.8	5.8	6.8	7.3	9.1	11.1	13.0
	Finance	1.7	2.1	2.5	2.9	4.5	5.7	7.1
	Real Estate	1.9	1.6	2.0	2.5	2.8	3.5	4.8
	Tenancy and Commercial Services	0.3	0.4	0.7	0.9	1.3	1.9	2.4
	Scientific Redearch, Technical Services and Geological Prospecting	0.7	1.0	1.2	1.5	2.0	2.9	3.7
	Other Industries	5.8	5.1	5.9	6.8	8.5	11.1	12.4
	Finance as % GDP	**4.4%**	**4.4%**	**4.3%**	**4.3%**	**5.5%**	**5.5%**	**6.1%**
	Finance as % Service GDP	**9.6%**	**10.4%**	**10.4%**	**10.6%**	**12.8%**	**12.8%**	**13.5%**

Source: Same as Figure 7.3

*Note: Figures in this chart are converted from Chinese Yuan (RMB) at the current exchange rate of 6.45 RMB = one U.S. dollar.

almost identical industrial structures in 1978, Beijing and Tianjin followed distinctive trajectories from 1979 in the relative rise of services and the decline of manufacturing. Although appearing to move in the direction of Beijing's service sector in 2002, Tianjin's manufacturing and service sectors began to diverge and reached the widest margin in the postreform era by 2008 due to TBNA's growing manufacturing contribution to the city's economy. In the meantime, the divergence between the manufacturing and service industries of Beijing became the widest by 2009 (Figure 7.3).

In trying to escape Beijing's huge shadow, Tianjin and TBNA began to compete with Beijing in favoring tertiary industry, particularly financial services. In 2009, Tianjin secured national support to become the financial center of northern China, while TBNA would serve as the hub of "innovative financial companies." At the same time, a document published by the Beijing municipal government set the goal for the capital city to become a "national financial center for decisions, management, information and service with global influence."[5]

China does not need two financial centers so close to each other, and the chance for TBNA to win this race is rather slim (Zhu 2010). Most headquarters of Chinese banks, along with headquarters and offices of multinational investment companies, are located in Beijing. An already established financial sector contributes to 13 percent of Beijing's GDP, doubling the contribution of the financial sector to Tianjin's GDP. However, the financial sector's share of the service sector's GDP for Tianjin rose steadily as its manufacturing sector continued to grow at the expense of services (see Table 7.1). In a related manner, TBNA faces difficulty in attracting a highly skilled labor force. In 2006, only 1.2 percent of the employees in Tianjin came from outside the city, excluding those who had attended colleges locally (OECD 2009). Many graduates of Tianjin's universities would choose Beijing over Tianjin, creating a competitive disadvantage for TBNA against its powerful neighbor in developing high-end services.

In envisioning and supporting the development of financial services and R&D facilities in TBNA, the central government has been partially blind to the large shadow of its home city. Even if TBNA can close some of the gap with Beijing due to powerful central support, as the data on 2009 suggest, the competition would still favor the side with the initial advantages and render the "Beijing effect" on TBNA more negative over time. It is incumbent upon the authorities of the national government and of Beijing and Tianjin to fully evaluate and balance the double-sided "Beijing effect" in planning and guiding TBNA's development.

Local Autonomy and Limitation

Seen from the global city perspective, linkages with the global economic network can free a city from its geographic boundary and control of the nation-state involved. Looking at TBNA through the global city-region lens,

Table 7.2 Gross Domestic Product (GDP) and Foreign Direct Investment (FDI) in TBNA and Tianjin, 2004–2009

	GDP of TBNA (Billion U.S. Dollars*)	% of Tianjin's Total GDP	Contracted FDI (Billion U.S. Dollars*)	% of Tianjin's Total FDI	Realized FDI (Billion U.S. Dollars*)	% of Tianjin's Total GDP
2004	19.4	42.7%	3.8	67.3%	1.7	70.6%
2005	24.9	43.9%	5.0	68.1%	2.5	76.0%
2006	30.4	45.0%	6.2	76.2%	3.3	80.9%
2007	36.7	46.8%	7.7	66.7%	3.9	74.2%
2008	48.1	48.8%	9.2	69.1%	5.1	68.5%
2009	59.1	50.7%	10.9	75.8%	5.8	63.9%

Source: Same as Figure 7.1

*Note: Figures in these columns are converted from Chinese Yuan (RMB) at of one USD.

there is the question of how TBNA relates to Tianjin as it becomes more global and regional in the Jing-Jin-Ji context. As Table 7.2 shows, TBNA accounted for half of Tianjin's GDP in 2009, whereas its contracted foreign investment as a proportion of Tianjin's total stood at 76 percent. Besides functioning as the major engine of Tianjin's economic growth, TBNA is a powerful vacuum that has sucked in and away more foreign capital that might have otherwise spread across the Tianjin municipal territory.

A dominant economic player for Tianjin aside, TBNA was criticized as having an ineffective administrative structure and cumbersome relationship with Tianjin (Xiao 2006; Zhu and Sun 2009). The separate districts that comprised the TBNA territory had altogether seven police bureaus, five business bureaus, and thirteen different bureaus in charge of taxation. This heavy and internally competing bureaucratic setup had sustained slow decision making, duplicative functions, and many superfluous party and government positions. In November 2009, the central government approved Tianjin's reform plan to eliminate three separate administrative districts and merge them into one single government—Tianjin Binhai New District—with a more simplified and unified governance structure. This new district administration was expected to have greater autonomy in making decisions and implementing further administrative reforms. *The People's Daily*, the official national newspaper, quoted a claim of Tianjin that it would release the power to TBNA so that it could make any policy decisions that used to be made by the municipal government, and would let TBNA to report to the central government directly if the decision has to be vetted by Beijing.[6] For example, the central government would preapprove the change of land use for a five-year period, instead of once a year, and allow the local government to allocate the approved land at its own discretion.[7] However, Beijing did not forget to install a check on TBNA's autonomy by insisting on a "one permission, one reform" policy requiring the new district government to ask for permission for every new experiment they would like to try.

The director of the new TBNA administration also serves the city of Tianjin as its vice party secretary, which keeps him within the local political pyramid." However, this position is high enough to be directly appointed by the central government. The first new director, He Lifeng, was the former mayor of Xiamen, a key city facing Taiwan in southeastern China. That he has never been an official of Tianjin indicates his appointment as a political strategy of the central government to put TBNA under its direct control. This signifies that TBNA's local autonomy does not translate into autonomous policy action when it is geographically and administratively so close to the large shadow of Beijing, both as a more powerful competing city and as the site for one of the most powerful central governments in the world.

Regional Opportunity vs. Constraint

To the extent that the comparative advantages and local autonomy of TBNA have facilitated and will continue to foster its dynamic growth, the "Beijing effect" and central government monitoring tend to act as constraints. The balance between these factors will depend on the critical middle and bridging Jing-Jin-Ji region that provides additional opportunities and constraints. To understand how this broader regional effect works, we subject TBNA in the Jing-Jin-Ji to a limited comparative analysis involving the Pearl River Delta (PRD) and the Yangtze River Delta (YRD) that surged in the 1980s and the 1990s, respectively (Chen 2007).

Located in southern China at the mouth of the Pearl River, the PRD is subject to varied definition of its spatial boundary. Chinese official documents often confine it to a portion of Guangdong province, whereas scholars tend to include Hong Kong and Macao in the PRD. There is a lack of consensus regarding what comprises the region's core. If we include Hong Kong, then its global connections and economic power make it the indisputable center. But if Hong Kong is not included, which does not make economic and geographic sense in our view, then Guangzhou, the capital of Guangdong province, will contend with Shenzhen for the status and influence of the PRD's center.

The region began to grow rapidly in the 1980s, mainly through absorbing a significant amount of overseas investment and specializing in export-oriented manufacturing. It started with a relocation of Hong Kong's factories across the border for the region's competitive low labor cost. Shenzhen, bordering on Hong Kong, was the first PRD city to take off due to its SEZ status. As Shenzhen industrialized further and became more expensive for manufacturing, Hong Kong capital spread out to the broader PRD to seek lower production costs until it became a dominant node of global manufacturing, especially for certain products. In 2001, 78.8 percent of China's telephone export originated from the region, as well as 34.8 percent of color TVs and 35.5 percent of VCD players. If the PRD were a country, it would be the sixteenth largest economy in the world and the tenth leading exporter (Chen 2007). The PRD now features other major manufacturing centers besides Shenzhen. In Dongguan, a secondary industrial center, a single plant can produce 60 percent of the electronic learning devices sold in the U.S. market, in addition to producing 30 percent of the magnetic recording heads used in hard drives worldwide (Gill and Kharas 2007).

The PRD has a great potential for regional economic integration, but it has not been fully pursued and achieved. This is partly due to multinational companies' interest in controlling the largely regionalized production chain. In Dongguan, for example, 95 percent of the parts and components of personal computer assembly can be found within Dongguan's jurisdiction, but foreign companies, especially those from Korea and Taiwan, "rely primarily on their transplanted supplier networks to minimize the use of local suppliers who are often perceived as cheaper, but less qualified" (Chen 2007), thus gaining a strong price squeeze at both ends. While becoming more blurred and porous over time, (inter)national borders can continue to be obstacles to regional cooperation (Chen 2005). Hong Kong and the rest of the region are divided by a border that remains administratively demarcated and controlled. Passengers heading for both sides of the Hong Kong–Guangdong border, on a daily or longer-term visit, still have to wait to pass the border checkpoint.

Hong Kong, if counted in the region, is undoubtedly the leading city of PRD. It has been the primary investor and job creator in the region. From 1979 to 1999, more than 40,000 Hong Kong–owned companies and factories in the PRD employed 10 million workers (Enright and Scott 2005). Hong Kong also is the center for industrial design and distribution, whereas the rest of the region is specialized in or confined to assembly-based production. A Hong Kong–based toy producer can create the designs, transport samples and likely raw materials to the factories, and distribute the finished products out of China through Hong Kong toward their destined markets (Chen 2007). Finally, Hong Kong is the service center for the region as well, based on its entrenched and irreplaceable tradition and foundation in advanced and concentrated financial services, including multinational banks and insurance companies (Yeh 2005).

Shenzhen has gained a prominent position within the PRD region. Set up as an SEZ in the 1980s to attract initial investment from Hong Kong, Shenzhen has grown tremendously and risen from a small town to a megacity of over 12 million people in less than twenty years (Chen and de'Medici 2010). However, as Shenzhen is already a megacity with its own economic strength and regional functions, it is becoming less dependent on Hong Kong. GDP per capita of Shenzhen reached about $13,000 in 2008, the highest among all Chinese cities. With continued industrial upgrading to overcome its weakness in labor-intensive manufacturing and transition to more services, Shenzhen's industries have become more technology- and knowledge-intensive, and its service industry accounted for over 50 percent of its total GDP in 2008 (see Figure 7.2). Home to one of the only two stock exchanges in China and one of China's most dynamic commercial banks (China Merchant Bank), the Shenzhen government in 2009 announced an ambitious goal to develop into a "global financial service center."[8]

The YRD is located at the central section of China's coastal belt and anchored to Shanghai. It features a main corridor of Shanghai-Suzhou-Hangzhou and a radiating area that extends to large portions of Jiangsu and Zhejiang provinces and beyond. The YRD has been China's leading economic region since the late Ming Dynasty 500 years ago, and began to develop a dynamic commercial tradition in the late nineteenth century. However, the region was not the first to be open up to foreign investment after 1978 and fell behind the PRD for nearly a decade till the early 1990s, when Pudong was developed as an SEZ and opened to global investment.

If the PRD is heavily labor-intensive, the YRD has a wider range of industries, including capital- and technology-intensive industry such as automobile and notebook computers, alongside labor-intensive garment or textile factories (Lu 2009). These industries, however, are similar to those in the PRD in that they also are dependent on foreign capital. Quanta, a Taiwan-based notebook maker ranked first in the world, owns a $48 million factory complex in Shanghai which employs 20,000 workers and produces more than 90 percent of its total notebooks for Dell and HP's U.S. market. Whereas manufacturing, packaging, and shipping are done in and from Shanghai, the most valuable components of the notebooks are designed and sourced overseas (Chen 2007).

Shanghai is usually regarded as the center of the region. It has been so since the early twentieth century, when it was the "Paris of the East," and has largely reclaimed this dominant position with an annual growth of over 12 percent since early the 1990s (Chen 2009). The city has served the region as the gateway to the rest of the world with its port ranking first in terms of container shipment in China and now second in the world. It also provides the region with educated labor through its large number of excellent universities and research institutes. More and more multinationals have set up regional headquarters in Shanghai to integrate the marketing/service and manufacturing functions in and around the YRD. With the help of these firms, Shanghai is being transformed from a primarily industrial city to a growing service center providing the YRD with human talents in manufacturing, service, and increasingly R&D.

Positioned at the center of Shanghai, Pudong is often considered the jewel in the crown of the YRD. Lujiazui, the heart of Pudong, is frequently referred to as the "Manhattan of China" and the dominant center for the YRD's service economy and regional corporate headquarters. Another part of Pudong, Zhangjiang High Tech Park, has become the R&D center for the greater region, with the third highest input of R&D investment of China's localities and significant spillover effects (Liu 2007). Pudong also has a free-trade zone that is directly linked to the port.

However, the leading role of Pudong is challenged from several directions. A number of secondary ports, such as Taichang in Jiangsu province, have been upgrading their capacities in order to capture some of the huge manufactured exports in container shipments that mostly exit through the Shanghai port. Suzhou, a booming industrial center one hour away from Shanghai, has established two industrial districts that have attracted a soaring number of manufacturing and even R&D facilities, such as that of Samsung's semiconductor operations in China. The only aspect of Shanghai and Pudong that is likely to remain unchallenged is the financial-service sector in Lujiazui, which stems partly from Shanghai's heritage and tradition and partly from the central government's goal of making it the financial heart and hub of China.

This brief assessment of the PRD and the YRD places TBNA and the Jing-Jin-Ji region in a clearer comparative light. Unlike the other two cases, the Jing-Jin-Ji region features extremely uneven spatial development. Wang's (2009) three-dimensional model of GDP contour lines reveals this uneven growth and resource distribution (see Figure 7.4). The figure shows a dominant peak in Beijing, the nearby lower peak representing Tianjin across a

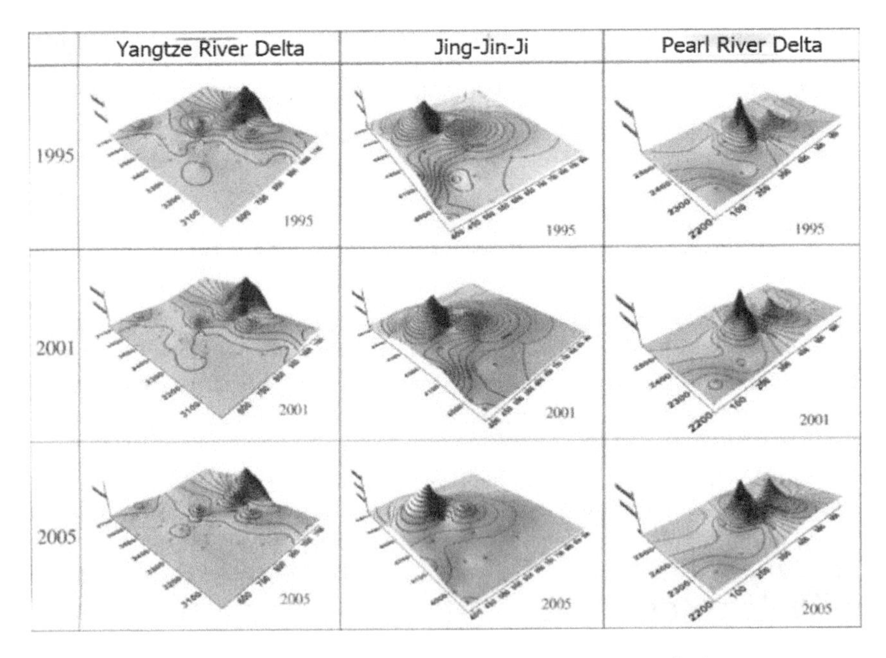

Figure 7.4 A GDP contour model of three major city-regions of China.
Sources: From Wang (2009: Figure 5), with the author's permission.

valley, and four or five scattered tiny hills on a nearly flat surface. In contrast, the figure displays more hills across the other two regions, especially the YRD, where powerful secondary industrial centers like Suzhou and Hangzhou have emerged. In the PRD case, we see a visible bridge connecting the two peaks of Guangzhou and Shenzhen, which is missing in the Jing-Jin-Ji case between Beijing and Tianjin. In statistical terms, the average per capita GDP of Hebei province's thirty-one counties around Beijing and Tianjin was $800, only about 16 percent that of Beijing (Jing et al. 2006). Unlike the PRD or the YRD, where a range of secondary and smaller cities compete with and are a complement to the central hub and one another for investment, the secondary and smaller cities in the Jing-Jin-Ji region are not sufficiently appealing to the global investment that flows automatically to Beijing first and to Tianjin, specifically now to TBNA, as a second destination, and they hardly participated in forming any value chain around the two cores.

The rise of TBNA has given the Jing-Jin-Ji region a new boost, but also poses a new challenge. While generally regarded as a crucial part of Tianjin, TBNA is gaining more economic and political autonomy vis-à-vis its host city, raising the question of whether it may become a quasi-independent city by itself. Given its national prominence, TBNA may see its relationship with Beijing as more important than with Tianjin. This reminds us of the relationship between Shenzhen and Hong Kong. For both pairs of cities that are secondary to the respectively dominant regional centers, the distance is very close, and the relative economic structures are comparable. TBNA and Shenzhen both started as manufacturing centers, and are now trying to step out of the shadow of their dominant neighbors by upgrading their manufacturing and by developing advanced services like finance.

Pudong is more often used as a reference case for TBNA in the sense that they are both special districts within a municipality, but Pudong has quite different regional functions than TBNA. First of all, Pudong is adjacent to downtown Shanghai, just across the Huangpu River, and thus much less likely to gain the semiautonomous status that TBNA may obtain. Pudong also has the most concentrated financial-service sector of any subcity district in China, which is unlikely to happen in TBNA in the foreseeable future. Pudong was the only district of Shanghai that was granted special policy incentives for luring banks and financial service firms. This is not the case with TBNA because Beijing, just one hour away, is already a well-established financial center, as discussed earlier.

These comparative evidence and perspectives reinforce the question about whether or not TBNA should be developed into a regional financial center. Instead, TBNA may be better off to follow the gradual upgrading path of Shenzhen in the PRD. Shenzhen developed with the help of Hong Kong's physical and intellectual capital and technology, and became fully industrialized while the labor and land costs grew. This prompted labor-intensive factories to migrate to the inner PRD. TBNA can help fill the largely empty economic valley of the Jing-Jin-Ji region by making full use of global capital and Beijing-based R&D facilities to strengthen industrialization further. At a future inflection point, industrial upgrading and saturation in TBNA will push less competitive and lower value-added manufacturers out to the surrounding areas. But because land supply is still plentiful in TBNA, the onset of this critical transition point is difficult to predict. Until that happens, uneven intraregional development may continue as much of the investment and resources are allocated to TBNA both through the top-down national promotion and through the horizontal inter-city competition that favors Beijing over the rest of the Jing-Jin-Ji region.

CONCLUSION: TBNA, REGIONAL GOVERNANCE, AND GLOBAL COMPETITIVENESS

TBNA is a timely case that enriches the theoretical debate and empirical study on the role of secondary cities and city-regions of developing countries in generating growth and distributing wealth as they become more globally connected. A. J. Scott (1996) argued that cities in developing countries have become "islands of prosperity" where global capital exploits the local surplus of cheap labor and resources. The more they are connected to the global market, the more they are exploited. Unless there are more integrated urban systems where complete value chains are locally and regionalized embedded, the larger share of the economic fruit will be collected by multinational corporations. This challenge is generally quite evident in China, but varies regionally and locally as shown through the comparative analysis featuring TBNA. Despite starting at a more industrialized stage as part of Tianjin than Shenzhen in the PRD, TBNA has not progressed far enough as a critical node of a regionally integrated economic system. Despite being the only assembler of Airbus passenger jets outside the developed world, the brand new facility in TBNA, visited by both authors of this chapter in summer 2009, imports about 90 percent of all the intermediate inputs and basic components. In other industries, TBNA and the Jing-Jin-Ji region lag behind the most advanced YRD and even the PRD in developing a regionally integrated supply, manufacturing, and distribution system.

This set of comparative evidence circles back to the theoretical discussion at the beginning of this chapter. From the familiar economic angle, a global city-region gains its competitiveness through economies of scale, both internal and external. Internal economies of scale can be obtained with increased production ability of single firms or industry in a broader sense, whereas the external can only by obtained when a complementary set of industries producing positive externalities concentrated in a region. Because a value chain is a type of external scale economy, well-developed transport infrastructure, which lowers transaction costs, can make a value chain more regionally rooted and thus globally more competitive. But the effects of sociocultural variables such as tradition and institutions advocated by Storper (2009) are less clear in the so-called "black box" of regional development (Florida, Mellander, and Stolarick 2007). Besides the obvious absence of tradition and institutions such as local knowledge in a new SEZ like TBNA, it embodies a powerful external context at a regional and national scale that both shrouds and illuminates the "black box" that is the Jing-Jin-Ji. One is tempted to see TBNA as merely a repeated product of central government policy like Shenzhen and Pudong. True as it is regarding the rationale and mechanism used by the national government to launch TBNA, this convenient political explanation would leave the various elements of the "black box"—a distinctive regional context of the Jing-Jin-Ji for TBNA—unexplained. We have dealt with this analytical challenge via a comparative approach to the timing and stage of city-regional development beyond initial creation.

In the PRD, which has a longer postreform development history behind it, the less valuable parts and components of the main products can be sourced within the region, but such integration has not been fully materialized due to the vested interests and advantages of global capital. Despite taking off later than the PRD, the YRD has caught up and moved ahead due to its distinctive advantages: a stronger and richer industrial tradition, coupled with Shanghai's integrating role and a density of productive secondary cities, have facilitated a more regionally linked economic system with broader spatial spillovers and less polarization and fragmentation. For TBNA and the Jing-Jin-Ji region, however, development is still very early and dominated by the overwhelming advantage of Beijing and the close, albeit delicate, relationship between Beijing and Tianjin. On the other hand, this early development, if well planned and with central government guidance, may help avoid the pitfalls the other two regions have differentially experienced, such as an overemphasis on export-oriented manufacturing that has delayed industrial upgrading, especially in the PRD (Chen 2007).

To the extent that TBNA is the third time that the national government has prominently and powerfully promoted a single area for focused and intensive development of the highest priority, it highlights broad regional governance as key to the success of this strategy. The National Development and Reform Commission (NDRC), a key central government agency, has attempted a comprehensive plan for the Jing-Jin-Ji region since 2007. This plan, however, has not worked out due to the difficulty in balancing the respective interests of the three provincial units that make up the region. In an ideal division of labor, Beijing would be specialized in the technological and financial services, TBNA in higher value-added manufacturing, and the secondary cities in lower-end production. But so far, TBNA has absorbed the bulk of new manufacturing investment while trying to grow advanced services, which leads to uneven regional development and sustains losing competition with Beijing.

As a final comparative reflection, TBNA in the Jing-Jin-Ji region has some similarity to the role and relational influence of Shenzhen in the PRD and Pudong in the YRD, or any hub, artificially created or "naturally" evolved, in any city-region for that matter. As TBNA develops further, it becomes more intertwined with its *local* and *regional contexts* and less shaped by the powerful initial impact of central government policy. The interaction between the opportunities of fast growth and greater autonomy and constraints of the "Beijing effect" and uneven development will intensify, and the outcome will depend heavily on instituting an effective governance mechanism of intraregional negotiation and cooperation to foster broader regional growth and stronger global competitiveness.

As key policy elements of this regional governance, the new TBNA government should better coordinate its development with Tianjin and other cities in the region instead of only focusing on and basking in its own spectacular growth in recent years. It will make sense for TBNA to concentrate on high-tech manufacturing and to tie up with the financial and R&D services in Beijing, instead of trying to develop them independently. Relative to governmentally "forced" or directed upgrading from manufacturing to services, a more evolutionary upgrading of TBNA's economy appears to be the best way forward. As the third coming of China's SEZs, TBNA has achieved the same spectacular growth as the first two rounds with Shenzhen and Pudong. As we end on the policy implications of our analysis, we also leave you with the caution that this chapter has shed only limited light on this complex model of development. Future research following TBNA's unfolding trajectory will surely yield more insights into the many facets and variations of secondary cities in the emerging global city-regions of the twenty-first century.

ACKNOWLEDGEMENTS

Chang Liu's field research in Tianjin Binhai New Area (TBNA) in summer 2010 that contributed to this chapter was supported by a student grant from The Kenneth S. Grossman '78 Global Studies Fund from the Center for Urban and Global Studies at Trinity College. Xiangming Chen would like to acknowledge the Paul E. Raether Distinguished Professorship for partially funding his research trip to TBNA with Chang Liu in summer 2010.

NOTES

1. State Council of the P.R.C. 2006. Opinions on the Development of Tianjin Binhai New Area, May 26; retrieved on November 5, 2009, from http://www.tjcac.gov.cn/binhai/detail.asp?articleid=9701&classid=268&parentid=0.
2. Zhang, Yongzhi, 2009. "Tianjin Port Ranked Third in China." *Bohai Morning News*, March 12; retrieved on November 5, 2009, from http://www.bh.gov.cn/bhsh/system/2009/03/12/010023273.shtml.

3. www.bh.gov.cn. 2009. "Ten Campaigns of Binhai, Total Investment Exceeds 1.5 trillion Yuan," August 13; retrieved on November 5, 2009, from http://www.bh.gov.cn/zjbh/system/2009/08/13/010031871.shtml.
4. Tianjin Municipal Commission of Development and Reform. 2009. "Plan to Advance Innovation Financial Services in TBNA is Approved," October 28; retrieved on January 12, 2010, from http://www.tjdpc.gov.cn/templet/default/ShowArticle.jsp?id=12798.
5. Beijing Official Website International. 2008. "A Plan to Advance the Financial Industry in Beijing," April 30; retrieved on January 12, 2010, from http://zhengwu.beijing.gov.cn/gzdt/gggs/t968340.htm.
6. www.peoples.com. 2009. "Unification of Administration in TBNA," November 10. Retrieved on January 12, 2010, from http://unn.people.com.cn/GB/14748/10346910.html.
7. See *Development Report of Tianjin Binhai New Area*, published by Administration of Tianjin Binhai New Area, 2008, pp. 46–47.
8. Finance Net. 2009. "Shenzhen Failed to Become a Global Financial Center," May 27; retrieved on February 28, 2010, from http://www.caijing.com.cn/2009-05-27/110172330_1.html.

REFERENCES

An, Husen, and Fei Chen (2009). "Barriers of Human Capital Flow to Tianjin Binhai New Area." *Science of Science and Management of Science and Technology* 30 (7): 191–196.

Arrow, Kenneth (1962). "The Economic Implications of Learning-by-Doing." *Review of Economic Studies* 29: 155–173.

Bai, Zhonglin, and Jun Li (2003). "A Compare Study of the Structural Features and the Changes of Three Industries in the New Districts of Tianjin-Binhai and Shanghai-Pudong." *Science of Science and Management of Science and Technology* 24 (7): 29–52.

Brito, Carlos, and Ricardo Correia (2010). *Regions as Networks: Towards a Conceptual Framework of Territorial Dynamics.* Working paper, Faculdade de Economia, Universidade do Porto.

Chang, Xinghua (2007). "SWOT Analysis of Jing-Jin-Ji Metropolitan Region." *Reference for Research on Economics* 2064: 28–37.

Chen, Xiangming (2005). *As Borders Bend: Transnational Spaces on the Pacific Rim.* Lanham, MD: Rowman & Littlefield Publishers.

Chen, Xiangming (2007). "A Tale of Two Regions in China: Rapid Economic Development and Slow Industrial Upgrading in the Pearl River and the Yangtze River Deltas." *International Journal of Comparative Sociology* 48 (2–3): 167–201.

Chen, Xiangming, ed. (2009). *Shanghai Rising: State Power and Local Transformations in a Global Megacity.* Minneapolis: University of Minnesota Press.

Chen, Xiangming, and Tomás de'Medici (2010). "The "Instant City" Coming of Age: The Production of Spaces in China's Shenzhen Special Economic Zone." *Urban Geography* 31 (8): 1141–1147.

Enright, Michael J., and E. E. Scott (2005). "China's Quiet Powerhouse." *Far Eastern Economic Review* 168 (5): 27–34.

Florida, Richard, Charlotta Mellander, and Kevin Stolarick (2007). "Inside the Black Box of Regional Development: Human Capital, the Creative Class and Tolerance." Working paper, the Martin Prosperity Institute, Joseph L. Rotman School of Management, University of Toronto.

Florida, Richard, Tim Gulden, and Charlotta Mellander (2007). "The Rise of the Mega Region." Working paper, the Martin Prosperity Institute, Joseph L. Rotman School of Management, University of Toronto.

Gill, Indermit, and Homi Kharas (2007). *An East Asian Renaissance: Ideas for Economic Growth*. Washington, DC: World Bank.

Jing, Tihua, et al. (2006). *Report on Regional Economies of China*. Beijing: Social Sciences Documents Press.

Lei, Ming (2007). "Advantages and Disadvantages of Tianjin Binhai New Area and the Future of Pan-Yellow Sea Cooperation." *City* 7: 17–21.

Li, Bing (2007). "Opening and Development of Tianjin Binhai New Area and the Cooperational Growth of Northeast Asia." *Hong Guan Jing Ji Guan Li* 10: 56–57.

Li, Jinkun (2008). *Growth and Future of Tianjin Binhai New Area*. Tianjin: Tianjin Academy of Social Science Press.

Li, Xuemei, Zhen Zhang, Xiangshu Yao, and Qianjin Zhang (2009). "Regional Assessment of Land Intensive Utilization for Tianjin Binhai New Area." *Journal of Shanxi Agriculture University* (Social Science Edition) 8 (3): 293–298.

Liu, Gongye. 2008. *Third Pole: Record of Development of Tianjin Binhai New Area*. Tianjin: Tianjin People's Publishing House.

Liu, Yong (2007). "On R&D Spillover within Yangtze River Delta and Economic Development." *Journal of Nantong University: Social Sciences Edition* 23 (2): 36–39.

Lu, Mingyuan, and Kai Liu (2009). "Comparison of Evolution of Industrial Structure between the Binhai New Area in Tianjin and the Pudong New Area in Shanghai." *Journal of Tianjin University of Commerce* 29 (3): 31–35.

Marshall, Alfred (1920). *Principles of Economics*, London: Macmillan.

Meng, Guangwen (2003). *The Theory and Practice of Free Economic Zones: A Case Study of Tianjin, People's Republic of China*. Doctorial dissertation submitted to Ruprecht-Karls University of Heidelberg, Germany.

OECD (2009). *Trans-Border Urban Co-operation in the Pan Yellow Sea Region*. Paris: OECD Publishing.

Ren, Yuan, Xiangming Chen, and Dieter Läpple, eds. (2009). *The Era of Global City-Regions* (in Chinese). Shanghai: Fudan University Press.

Romer, Paul M. (1990). "Endogenous Technological Change." *Journal of Political Economy* 98 (5): S1071–1102.

Sassen, Saskia (1991). *The Global City: New York, London, Tokyo*. Princeton, NJ: Princeton University Press.

Scott, Allen J. (1996). "Regional Motors of the Global Economy." *Futures* 28: 391–411.

Storper, Michael (2009). "Regional Context and Global Trade." *Economic Geography* 85 (1): 1–21.

Vogel, Ronald K. (2010). "Governing Global City Regions in China and the West." *Progress in Planning* 73: 4–10.

Wang, Wei (2009). "A Comparative Study of Eco-Spatial Morphological Features of Three Major Urban Agglomerations in China." *Urban Planning Forum* 179: 46–53.

Wu, Liangyong (2005). "Assessing the Strategic Significance of Tianjin Binhai New Area from Jing-Jin-Ji and Broader Spatial Context." *Port Economy* 5: 5–7.

Wu, Zhe (2007). "The Future of Jing-Jin-Ji Region and the Aspects of Zhongguancun, Caofeidian and Binhai New Area." *China Reform* 10: 44–46.

Xiao, Jincheng, Yulong Shi, and Zhong Li (2006). *The Rise of the Third Growth Pole: Research on Development Strategy of Tianjin Binhai New Area*. Beijing: Economic Science Press.

Yeh, Anthony G. O. (2005). "Producer Services and Industrial Linkages in the Hong Kong- Pearl River Delta Region." Pp. 150–173 in *Services Industries and Asia-Pacific Cities*, edited by P. W. Daniels, K. C. Ho, and T. A. Hutton. London: Routledge.
Zhu, Erjuan (2010). *Studies on Industrial Cooperation between Tianjin Binhai New Area and Beijing*. Beijing: China Economic Publishing House.
Zhu, Xufeng, and Bing Sun (2009). "Tianjin Binhai New Area: A Case Study of Multi-Level Streams Model of Chinese Decision-Making." *Journal of Chinese Political Science* 2009: 191–211.

8 Social Accountability in African Cities

Comparing Participatory Budgeting in Johannesburg and Harare[1]

Tyanai Masiya

INTRODUCTION

Globalization has become an increasingly powerful force that affects all socioeconomic systems and political life at both national and local levels around the world. Urban development and city growth, which has been accelerating in Africa, has been caught up in the simultaneously complementary and contentious nexus between global influence and local traditions. From a global perspective, the ideas and practices in urban development related to planning, design, and governance have crossed national and city boundaries and triggered imitative policies and practices (Zukin, this volume), whereas entrenched local conditions can continue to hamper them. Today many African cities are adopting participatory budgeting in a manner aligned with a larger global trend. Participatory budgeting is a social accountability mechanism that has its origins in the city of Porto Alegre, Brazil, but has since spread to Europe, Asia, and Africa over the last two decades or so (Sintomer et al. 2010: 9).

In theory, participatory budgeting supports the deployment of development resources through the full participation of potential recipients of the development programs (Langa and Jerome 2004: 3). It thus provides a framework that encourages civic participation at and from the community or grassroots level. Participatory budgeting also situates decision makers closer to those to whom they are accountable, discourages corruption and encourages efficiency, cooperation, and responsiveness. Participatory budgeting can help and empower a government that adopts it to protect precious resources and allocate them more equitably.

The adoption of participatory budgeting in African cities has followed a serious crisis of governance. In this context it is an indictment of the all too common top-down decision-making structure that vests limited power in local leaders. In the top-down decision-making process of the authoritarian or semiauthoritarian African countries, local leaders are largely accountable to the state, and not to the very constituencies they presume to serve. In the provision of services in African cities, this structure has often been unresponsive, inept, and corrupt. Hence the concentration of power at the central government level became a barrier to providing relevant and adequate resources and services to desperate and needy local urbanites. The result has been popular disillusionment with urban authority due to poor service delivery such as garbage collection and disposal. As a consequence of poor service delivery, typical features of African cities include overcrowding, lack of clean water supplies, poor waste-management services, and sprouting slums where poverty and pestilence are commonplace (UNEP 2002).

To address these challenges, a number of African cities have attempted to adopt social accountability mechanisms that enable local citizens to bring their leadership to prioritize and be accountable to citizens' demands in planning processes. This movement is part of a larger trend toward a greater role of civil society in increasing state accountability for development in Africa (Devarajan, Khemani, and Walton 2011). The success of participatory budgeting in Porto Alegre and other places has made it an appealing policy for African cities seeking political and other reforms. Indeed, the success and sustainability of urban development hinges on the implementation of reforms that engender popular empowerment and encourage civic participation. The involvement of citizens in participatory budgeting can promote consensus, accountability, and transparency in resource utilization. To the extent that social accountability mechanisms such as participatory budgeting are likely to achieve these goals, they have enjoyed growing support as a strategy for improving the management of African cities. This research traces the implementation of participatory budgeting in two African cities—Johannesburg (South Africa) and Harare (Zimbabwe), respectively—for their contrasting records on the process and outcomes of the implementation. For each city, the perceived and real consequences of participatory budgeting are examined, and the future challenges and prospects of its continued implementation addressed.

MOTIVATION BEHIND PARTICIPATORY BUDGETING IN AFRICAN CITIES

The adoption of participatory budgeting in many African cities, including Johannesburg and Harare, is a response to service-delivery challenges that they have faced. There is almost a universal imbalance between the resource support for supply of services and popular demand for them. This has sustained pervasive poverty, inequitable distribution of wealth, widespread mismanagement and corruption, all of which have contributed to a persistent urban crisis (Sten and Halfani 2001). Whereas different symptoms can be identified for different cities, service-delivery failure has been the common denominator of serious challenges confronting all African cities. A few examples suffice to illustrate this broad malaise.

Observing events in Swaziland, Wunsch (1998) argued that Swaziland's two main cities, Mbabane and Manzini, were plagued by poor utilization of revenue and service-delivery systems. In both cities, locally powerful individuals used their political connections to frustrate local tax-collection endeavors. Audits were years behind schedule. In South Africa and Tanzania, some municipalities were unable to collect taxes from citizens who resorted to evasion because they simply had no confidence in the local government's ability to properly use tax revenue for providing promised services, given the extensive corruption among local officials (Fjeldstad 2004, 2006; Jameson and Martinez-Vazquez 2003). In Harare, the capital city of Zimbabwe, the chairperson of the Combined Harare Residents Association (CHRA) and thousands of other residents refused to pay rates for two years in protest against poor service delivery and in opposition to the imposition of an unelected council (Kamete 2006). The Human Rights Watch (2007) reported that corruption in Nigeria's Rivers state region municipal governments was "so pervasive and so debilitating that, with the exception of paying civil service salaries, [local authorities] have virtually ceased to perform any of the duties assigned to them." Finally, a study carried out in Uganda (Rienikka and Svensson 2004) found that the bulk of the local school grants transferred from central government was seized by local government officials and politicians by force.

These examples support Mhlahlo's (2007) contention that a bureaucratic maze of nontransparent municipal administrative practices hampers service delivery and marginalizes the poor. He further argued that any improvement in the quality of urban governance would transform decaying, impoverished, and socially exclusionary slums into thriving cities characterized by prosperity and a civic spirit of inclusivity. This view anticipates and reflects a broader acceptance of participatory budgeting in urban Africa as a key mechanism for improving transparency and good urban governance practices.

PARTICIPATORY BUDGETING IN THEORY

In 1989, the city of Porto Alegre in Brazil embraced reforms that are now classified as participatory budgeting. By involving the citizens directly in the formulation and execution of the city's budget, participatory budgeting revitalized civic spirit and spurred sustainable development that marks Porto Alegre as a model for struggling cities of developing countries across the globe. This success story led to a rapid expansion of participatory budgeting reforms not only among Brazilian cities but also to urban communities from Latin America to Africa.

The theoretical rationale behind participatory budgeting postulates that truly democratic and transparent administration of resources is one of the most effective ways to promote urban development. Marquaetti (2007: 1) argued that under participatory budgeting, "*the citizens . . . have the right of participating in the elaboration of the fiscal policy of their municipalities, taking part on the definition of how and where the resources will be obtained and how and where they will be employed in their neighborhood.*" Such involvement can ensure that resources are allocated according to the demands of an enfranchised public and not consumed or wasted by the municipal authorities.

The difference between participatory budgeting and traditional budgeting is that the latter is a preserve of municipal officials led by a treasurer and director of finance whilst the former is a comanaged process that requires civic participation at every step in the process of crafting, implementing, and executing local budgetary measures. Shah (2007) noted that participatory budgeting helps citizens understand their civil rights, and encourages them to freely and openly express their opinions. When budgetary measures are implemented, participants are provided with proof that their views and actions indeed affect public policy. As citizens become more proficient in the art of public policy, their ability to hold municipal government accountable is augmented accordingly. The logical conclusion, then, is that governance will be more responsive and effective under the participatory budgeting process (Putnam 1993).

By and large, the strength of participatory budgeting is that it binds together the interests of the public and private sectors as well as city managers. This fosters social harmony, legitimizes urban governance, and creates a platform upon which economic recovery and sustainable development can be balanced (Wampler 2007: 21). Schugurensky (2004) identified participatory budgeting as an effective means for the development of infrastructure, which is very weak in many African cities. Hence it is suggested that with participatory budgeting, waste-management systems will improve, schools will receive badly needed resources, and cityscapes would be transformed from filthy slums to tidy, modestly constructed spaces that reflect the confluence of resource supply and demand that is required for even the most modest urban development.

There are distinct stages in participatory budgeting for an adopting city. It features a joint process between the city and its citizens through four stages. The first stage is preparatory, which ideally involves the distribution of information by city authorities, the creation of relevant participatory structures as well as the initial discussion or brainstorming of policies and priorities by citizens. The second step focuses on formulation where the mayor and his/her department leaders define the purposes, objectives, programs, operational details, expected local impacts, projected roles of council members, and prerequisite measures relating to the adoption of participatory budgeting. This stage also includes meetings for situational analysis for the purpose of identifying specific problems and needs. Citizens and their elected officials would then cooperate to set priorities and decide on project priorities. Their participation results in a draft budget that reflects the needs of an informed and participatory public that can be submitted to the full council by the mayor. The third stage involves implementing the budget, in line with the resources allocated during the formulation stage. In keeping with the ideals of transparency and cooperation, all activities directed to this end involve consultations between the municipal staff and a participatory budgeting committee. The participatory budgeting committee is created at the formulation phase and includes citizen and political representatives. At the final stage, the participatory budgeting committee provides community oversight and an impartial auditing authority. Councilors and citizens share the right to monitor budget implementation and collaborate when conducting onsite monitoring and project evaluation. These oversights produce performance reports that the municipality gives to citizens at regular intervals. This multistaged approach is supposed to promote accountability, transparency and sustainable citizen-centered development. The intended and actual outcomes of participatory budgeting will be evaluated in terms of what has happened in the cities of Johannesburg and Harare.

PARTICIPATORY BUDGETING IN JOHANNESBURG

Johannesburg is South Africa's largest city, with a population of about 3,200,000. The city's participatory budgeting process is institutionalized through the country's constitution, which recognizes the framework of local governance. Section 52 of the 1996 constitution defined local government as the custodian of democratic and accountable government for local communities. In terms of this section, local governments are required to provide democratic and accountable government, to provide services to communities in a sustainable manner, to provide social and economic development, to promote a safe and healthy environment, and to encourage the involvement of the community and community organizations in the matters of local government (Nyalunga 2007). Section 153 (1) of the constitution also outlines key duties of municipal governance as structuring and managing its administration and budgeting and planning processes to give priority to the basic needs of the community and to promote the social and economic development of the community (Yusuf 2004). Municipalities are also encouraged to involve communities and community-based organizations in their planning and operations. South Africa's legislative provisions that support participatory budgeting are also framed within the context of the country's constitution, specifically Chapter 7 on local government. These are the Local Government Act, Local Government Municipal Structures Act (1998), the Municipal Systems Act (2000), Municipal Finance Management Act (2003) and the Municipal Property Rates Act (2004), all of which ensure that participation is accessible to all, free, and fair (Yvette 2007), at least in theory.

The Municipal Finance Management Act (2003) provides the framework within which municipalities draw up budgets, with specific timelines for preparation and approval. The Municipal Property Rates Act (2004) also promotes community participation in the budget process and the determination of rate policies. The Municipal Structures Act (1998) requires that municipalities set up ward committees (smallest unit in the city's governance structure) to support effective participation at the local level. Together with the Municipal Systems Act (2000), the two acts provide for external consultations through both formal and informal means, including public meetings with residents, business, state departments, and similar entities, which is a core value of participatory budgeting. The participatory budget process in the city is funded through the aptly titled "budget office."

The process begins with the first mayoral Lekgotla (a local name for a meeting) that is conducted before the budget is drawn up. The Lekgotla evaluates the previous financial year's performance and plans for the future. Taking stock of the previous year's activities is a critical exercise that determines the development direction in which the city will proceed. Budget panel meetings follow the first Lekgotla. In the panel meetings, city departments and municipal entities present and assess proposals. The objective of the panel meetings is to ensure that the forwarded proposals are congruent with the provisions of the developmental objectives expressed in the city's Growth and Development Strategy (GDS) and to guarantee that the available resources are allocated in line with the city's priorities. The budget panel includes the city manager, chief financial officer, and the mayoral committee member for finance and economic development.

The panel's assessment report is presented at the second Lekgotla, which is held to assess the budgetary obligations of the next financial year. The two Lekgotlas enable planning to be more focused and detailed. The draft budget allocations are made and issued to the various departments and entities so that they may revise their budget proposals in line with the allocations. Through incorporating the outcome of the public participation process in the outreach phase, a consolidated draft budget is refined.

The outreach phase follows the two Lekgotlas. Participants in the outreach process include ordinary citizens, resident associations, community-based organizations, NGOs, religious groups, youth and women's formations, sports clubs, cultural groups, universities, technical colleges, trade unions and organized business. It is launched through an extensive communications campaign, designed to disseminate information on the participatory budgeting program and mobilize all stakeholders to participate in the activities. These include notices in the local newspapers, local radio stations, posters, leaflets sent out with invoices and the electronic media. The outreach phase is critical in that it promotes accountable and transparent governance, citizen ownership of projects, and demand-based project prioritizations. In order to enhance participation, knowledge and understanding, the needed participatory budgeting training for citizens is provided by NGOs and community-based organizations (CBOs). Many NGOs and CBOs provide free training within Johannesburg. The most prominent public interest organization providing such training is the Institute for Democracy in South Africa (IDASA).

The mayor's "road shows" (public rallies) are the strongest showing of the city's resolve to pursue participatory budgeting. These are intended to provide information and encourage participation. During these shows and other public meetings, the mayor is accompanied by heads of departments whose responsibility is to further explain budget procedures and the budget cycle; define entry points for citizen participation during the budget cycle; provide detail about local revenue sources, unpaid taxes and user charge arrears; and provide information on investment needs and actual investments.

In addition to meetings at the ward level (the smallest unit of a council structure), public meetings chaired by members of the mayoral committee and senior city officials are held in each region. These meetings further explore citizens' service-delivery expectations. At a city level, consultations are made with key stakeholder groupings, NGOs, CBOs, labor and women's groups. Such an extensive participatory process minimizes the possibility of misallocation of resources, breaks the cycle of exclusion of the marginalized, and promotes trust between the city government and its citizens.

Following the meetings, a synthesis of all the concerns and needs expressed in the outreach meetings is calibrated in the draft budget and again taken back to communities. The draft budget and all the requisite documentation are distributed in the wards and the regions and the community is walked through each of them. The city explains how it intends to address concerns raised through service-delivery targets for the ensuing year and proposed capital projects. Communities are also granted a platform to critique the way the city has elected to address their needs. A public participation report of the participatory budgeting process is prepared and forwarded to the mayoral committee.

After the mayoral committee meeting, the mayor presents the report before a full council for approval. Upon endorsement by a full council meeting, the city shares the approved budget with the provincial government. The new budget and tariffs are displayed in the city's Web site, public centers and promulgated in the government gazette. Following the coverage in local media, extensive publications keep the citizens informed and facilitate their participation during the implementation and supervisory stages of the budget. Problem-solving activities begin through respective departments and entities of the city throughout the financial year.

In addition, a monitoring process occurs during implementation. The city's management, councilors and ward committees regularly schedule meetings to appraise themselves about the developments within the prevailing budget. Evaluation is also part of the preparatory phase for the following year's budgeting process. Monitoring and evaluation include petitions. Petition procedures were designed to dispel any suspicions of political bias. A multiparty Petitions Management Committee is tasked by the mayor to consider complaints and give feedback to all petitioners within a reasonable time. Other measures include feedback on community radio stations, views posted on the city of Johannesburg Web site, and tips for the mayor are solicited in the printed press and in customer-satisfaction surveys. Such a level of transparency promotes whistle-blowing, minimizes corruption, embezzlement, and other forms of mismanagement.

Over the past five years, the city of Johannesburg also has carried out annual customer-satisfaction surveys. The surveys measured satisfaction in six broad categories. These six were basic services (including electricity, refuse collection, roads, sanitation, street lighting and water); community services (including ambulances, bus service, public toilets, litter removal, street sweeping, and traffic lights); public safety and bylaw enforcement services; billing, payments and enquiry services; communication and interaction efficiency; and corruption. The outcome of satisfaction surveys is important in that it measures the extent to which the city is responding to the needs of the citizens and guides future resource allocations according to the views of the citizens themselves. It allows for the government to obtain ground-level pulse readings.

As a result of the extensive participatory budgeting process, project implementation in the city of Johannesburg has been based on strong comments from the community, private sector, and NGOs. Participatory meetings at formulation, execution, and evaluation stages of the participatory budgeting process have been inclusive of all stakeholders. This has helped shape a robust planning process whose project implementation is based on citizen priorities. As a result of extensive citizen participation, Johannesburg's participatory budgeting process has been viewed as one of the most progressive in African cities.

PARTICIPATORY BUDGETING IN HARARE

Harare is Zimbabwe's capital and largest city, with a population of approximately two million. Local governance is not recognized in the country's constitution, but is governed by the Urban Councils Act. The ministry responsible for urban governance has regulatory functions in relation to certain aspects of policy and management of the city. Through the Urban Councils Act, the city is required to consult the community during the budgetary process. In recent years, the city's citizens have also taken a keen interest in the way the city is run and resident associations have been formed. Together with other bodies representing commerce and industry, they give input to the budget process during participatory budgeting meetings. In fact, Harare adopted participatory budgeting following a huge number of objections to its annual budgets (Toriro n.d.). This scenario increased mistrust and tension between council and residents that characterized the beginning of every financial year. This is because citizens were not aware or informed of the rationale behind the new budgets, particularly insofar as it applied to various increases in user charges and rates; hence their increased mistrust of city authorities. The city residents may become more trustful if the budget process can be made more transparent to reduce the possibility that deals are made behind closed doors.

Unlike Johannesburg, where training for participatory budgeting is more available, civil-society organizations (CSOs), including the Urban Councils Association of Zimbabwe (UCAZ), have occasionally conducted training for councilors, council officials, and the private sector to promote participation. The training activities create community-participation activities and stakeholders' knowledge about the municipal budget formulation, interpretation, review, implementation, and management process. Compared to Johannesburg, the participatory process in Harare is less complex but nevertheless consists of multiple stages.

City technocrats led by the city treasurer begin the budgetary process by collecting information, reviewing previous budget performance and the status of projects implemented the previous year. Councilors then bring project proposals from their wards, which are then incorporated into the budget proposals by the city's administrative departments. Upon completion of the draft budget, the council officials hold consultative meetings with the community, CSOs, and members of the business sector, reflecting the essential provisions in the proposed budget. Such a process is meant to vest ownership of the budget in the hands of the community. In practice, however, the city has not sufficiently consulted its citizens because of interference from the central government through the minister responsible for urban municipalities. This top-down approach reflects the authoritarian tendencies and practices of the Zimbabwean national government in recent years.

Pursuant to public consultations at the next step of the process, the council revises the draft budget in line with the feedback. The drafted budget is advertised over a month in advance, and in at least two issues of the national newspaper. During this period, residents and their associations, CSOs, and the business sector have the right to lodge objections to any provisions of the budget. The council addresses these objections either by providing a satisfactory explanation, amending the draft budget to cater for the objections or attending to any such matters raised as they pertain to the budget. Finally the city submits its proposed budget to the ministry responsible for urban governance, together with a certificate of proof of consultation for the ministry's perusal and final approval of rates. The certificate of proof of consultation is also accompanied by a detailed explanation from council on how it responded to all objections that may have been raised during the consultation process. Once approved, all services provided for in the budget are made available to the community. New rates, charges and fees are then effected while new projects in various phases are also implemented.

Once implementation begins, government officials, through the office of the town clerk, make periodic reports to council. These periodic reports are cascaded to the various wards and other stakeholders through councilors and the office of the mayor. However, the monitoring process for the city can be improved by further strengthening community participation. There is no deliberate effort to make evaluation a participatory exercise at the moment. Furthermore, numerous factors have militated against the smooth running of the participatory budgeting process in the city. The minister responsible for urban governance suspended the executive mayor elected on the opposition Movement for Democratic Change ticket in April 2003. He also suspended twenty of the twenty-one councilors, all from the same party. In terms of the legislative provisions, the minister may appoint a commission to run a city's affairs for not more than six months. The city of Harare was, however, run by a commission appointed by the minister from 2003 to 2008.

As a result of the political environment obtaining during this period, it was difficult for the city to provide for participatory budgeting. Its political leadership became accountable to the minister rather than the citizens. The budget had minimum if not insignificant citizen participation (Chaeruka and Sigauke 2007). For example, in its report on the 2004 Harare city budget, the Combined Harare Residents Association (CHRA) argued that the budget did not reflect the aspirations of the city's residents. The association pointed out that it was a result of an unacceptable process that did not seek comprehensive input from the residents, and was the product of technocrats within the municipality with minimal or no input from councilors who rubber-stamped it. In addition, during the budget notice period, the acting city treasurer instead of the councilors summarized two thousand five hundred objections to the budget.

Significant complaints were also raised pertaining to the city's failure to consult in the 2005, 2006, and 2007 budgets (Chaeruka and Sigauke 2007). The residents demonstrated and defied council officials who attempted to attend meetings to address the issues. On those occasions, the chairperson of the commission avoided attending the potentially charged meetings. For the 2008 budget, the commission made cosmetic consultations after it had already started implementing the budget (*Zimbabwe Gazette Daily News* 2008). Between 2001 and 2006, the city of Harare also failed to produce audited accounts of its books, even though the Urban Councils Act requires that the city produce such accounts annually and in a timely manner. Among other important functions, the audited statements inform the community as to how its money had been spent.

Selective implementation of participatory budgeting meant that service delivery continued to deteriorate. A highly unstable macroeconomic and hyperinflationary environment during the period exacerbated the matter. A budget crafted to last twelve months could last no more than three months, which made consultation on supplementary budgetary funding difficult. The city attempted to revive its fortunes by hiring a private consultant to craft its "Turn-Around-Strategy" (TAS) in 2006. The consultant worked privately and independently of the community and the city introduced a finished product for implementation. Harare then faced perennial shortages of clean water, disease outbreaks, drainage, sewage and waste-management problems, overcrowding, poor road maintenance, and high employee attrition. Corruption also reached an alarming level (*Transparency International, Zimbabwe* 2009).

In 2009, however, an elected council led by Muchadeyi Masunda scored a victory from the then opposition party in council elections and began to improve service delivery. Whereas residents have acknowledged improvement in delivering some services since then (Nleya 2011), there is still significant resistance to pay for services by residents, and the city relies on external funds such as provided by UNICEF to support its water purification.

CONTRASTING JOHANNESBURG AND HARARE

For both cities in this comparative study, constitutionally mandating municipal governance seems an important foundation for successful participatory budgeting. It minimizes interference from the national government and increases the autonomy of the cities. This in turn facilitates independent decision making and makes the leadership more accountable to their communities. In comparison, the city of Johannesburg is more successful in implementing participatory budgeting primarily because local governance is enshrined in the Constitution of South Africa. The city of Harare has been much less successful with participatory budgeting primarily because local governance is not written into the constitution, which renders the city less autonomous and more vulnerable to national government interference. Zimbabwe's minister responsible for local governance plays a supervisory role and has tended to interfere with the city's activities, including dismissal and suspension of mayors and councilors. Such a scenario promotes inefficiencies, creates a lack of transparency, and eliminates accountability to the constituent community. Consequently, most citizens have stayed from participating in the decision process as the city's leadership answers first to the ministry in charge.

A clear and unambiguous legislative framework would promote participatory budgeting. The pieces of legislation that govern Johannesburg give it a logical framework within which the city can promote comprehensive participatory budgeting and raise municipal finance. These pieces of legislation permit the city to draw up its budget and to promote community participation in the budgetary process and in influencing related policies. The legislation also outlines the structures that must be established at the local level to foster demand-based programming. Legislation that affords too much power to the parent ministry—as demonstrated in the case of Harare—tends to stifle participatory budgeting. As a result, the city of Harare operated without a democratically elected mayor and council from 2003 to 2008, following their respective suspensions by the supervising minister. The city also operated without substantive leadership of key departments including housing, treasury, public relations, and the office of the town clerk during this period. The commission running city affairs did not have a keen interest in comprehensive consultation processes, but instead served the interests of a minister who rewarded it with semiannual renewal of its tenure. The citizens were excluded, and their grievances regarding poor service delivery and resource allocation remained unaddressed.

Participatory budgeting is an expensive exercise. Funding for meetings and training citizens for full participation is an important requirement and a challenge for financially strapped cities. Once training has been provided and participation embraced, there must be sufficient funding to implement all prioritized projects and activities. Failure to provide such services can result in further self-exclusion, as citizens do not see the value of spending their time in the participatory process. The city of Johannesburg has an established budget office that is sufficiently funded to sustain citizen participation. The city also supports training activities to increase the capacity of citizens to participate in the budgetary process. It has formed a strong network of training institutions including universities and civic groups. This has helped citizens understand the logic of the political process, and their assumption of ownership of local programs is an important element for success. In underperforming urban economies like Harare, where capital projects ceased in the first decade of the new millennium, and the city was denied borrowing powers, the influence of participatory budgeting has been extremely limited. Citizens measure the success of participatory budgeting on the basis of capital projects. Furthermore, the lack of funding has meant that Harare has no resources to commit to a budgetary office. High inflation in fact rendered the participatory budgeting process meaningless—training is too expensive and participation is simply inconvenient. Therefore, participatory budgeting in Harare fared poorly when compared against Johannesburg.

The political environment is critical to determining the sustainability of the entire process. The success of the participatory budgeting process in the city of Johannesburg can also be attributed to the political will of its mayor, as well as other decision makers, including the speaker, the councilors, and administrators. In the city of Johannesburg, the mayor uses the political legitimacy of his office based on democratic election to ensure that participatory budgeting is always participatory and inherently transparent. For example, the city's 2008/9 participatory budgeting program was entitled *"Deepening Democracy, Enhancing Good Governance, and Building a Caring Society."* In practice, this initiative was a thorough, lengthy process that promoted participation and ensured what emerged as truly a "people's budget."

While Johannesburg has been able to deal with the demands brought about by multiparty political leaderships, which become more common as Africa democratizes further, Harare has been impaired by interparty conflict and state interference. The suspension of the mayor and councilors that occurred for years had no legal basis. It was blatant political repression of an opposition group by a dictatorial regime accountable only to itself. The minister responsible for local governments also dismissed elected officials and influenced the dismissal and appointment of handpicked heads of key administrative offices. Consequently, the mandate of the appointed commission was conferred from above, and its conduct reflected that arrangement between 2003 and 2008. While the democratically elected new Mayor Muchadeyi Ashton Masunda (a lawyer from the opposition to the national government) has since 2008 led to some improvement in urban governance and participatory budgeting, Harare continues to struggle in providing basic services such as water, waste removal, road repair, street lighting, and descent remuneration, leaving the residents, most of them poor, to fend for themselves regarding these basic municipal services.

The participatory budgeting process is best undertaken through scheduled public meetings, use of posters, pamphlets, and electronic and print media. Information dissemination must use all means available. The city of Johannesburg deliberately solicits support from CSO groups that have participatory budgeting expertise, and it reaches out to technical schools and universities, and other teaching and research institutions. In contrast to the city of Johannesburg, Harare's occasional budgetary meetings are only held in response to threats of court action, popular demonstrations, and refusal by residents to pay for inadequate and/or poor services. The ministry responsible for supervising urban governance had been at loggerheads with CSOs over providing local capacity building programs in Zimbabwe to such an extent that CSOs have been occasionally banned from conducting such services. When allowed to undertake training activities, the minister had compelled them to submit the training syllabi, objectives of the training, and its funding details for approval. Such interference with city affairs by the central government restricts local autonomy and curtails community participation, which is a key to achieving sustainable and equitable urban development.

Creating incentives for civil-society organizations and ordinary citizens to participate in policymaking is a necessary precondition for the success of participatory budgeting. Citizens need to have a sense of ownership in and commitment to the success of the process. Whereas participatory budgeting attracted huge enthusiasm in both cities in recent years, the number of citizens who are participating has been decreasing (Matovu and Mumvuma 2008). Reduced participation implies that citizens are not convinced of the changes that may be driven by participatory budgeting. In the case of Harare, it is clear that the leadership has not aggressively and honestly pursued participatory budgeting. The potential for civic participation in Harare has not been fully exploited. Self-interested national and local leaders continue to ignore repeated instances of remonstrance from civic groups, thereby refusing full commitment to participatory budgeting. In addition, the lack of transparency and continued exclusion of the constituent communities has maintained a political culture that is marked by desperation, division, and disillusionment.

CONCLUDING THOUGHTS

The participatory budgeting processes that African cities have embraced have seen mixed results. Where the process has been sufficiently implemented, as in the city of Johannesburg, participatory budgeting has helped make urban governance more transparent, accountable, and effective. With a strong and committed mayor at the helm, it has reduced some corruption and improved services. However, in cities like Harare that continue to face severe resource scarcity, legislative and political malfunctioning, and rampant corruption, efforts to implement participatory budgeting have been ineffectual and have thus hampered urban governance and development.

In general, participatory budgeting is a positive mechanism for development if it is implemented according to the principles of constitutional mandate, transparency, and true local community involvement. In addition, participatory budgeting in the two African cities exemplifies the transnational and translocal spread of ideas and practices in urban governance as cities become more globally connected (Kanna and Chen, introduction, this volume). Yet the two cities provide a sharp contrast in how the process and outcomes of implementing participatory budgeting can differ considerably depending on the national and local contexts, as well as the political commitment to making it work. This striking evidence points to the political or institutional barriers that can prevent even the best practices of urban governance adopted externally from indigenous success. This conclusion may be expected from the national contexts of the two chosen cities. As the capital city of authoritarian Zimbabwe, Harare pales in comparison to the more democratic system of postapartheid South Africa that has empowered the municipal government of Johannesburg to implement participatory budgeting.[2] Yet this is exactly the purpose of choosing two different cases for accentuating the great contrast in the process and outcome of adopting the same governance approach in two very divergent systems.

Finally, in contrasting Johannesburg and Harare, this chapter also sets up a distinction between the relative roles of the African state versus the Chinese state, both at the national and local level, in shaping the (mis)fortunes of their cities. Although the Chinese central government has given considerable autonomy to Tianjin and TBNA to grow their economies and attract foreign capital with top-down support (Chen and de'Medici, Liu and Chen, this volume), it will never go as far as allowing local government to experiment with participatory budgeting. The irony is that in spite of, or because of, its weaker status and capacity, the African state in both the South African and Zimbabwean contexts has introduced the more democratic municipal participatory budgeting, albeit with different results. This brief comparative reference also illustrates the very different domains for the Chinese and African states to engage the urban economy. Whereas the national and local governments of China have joined forces to drive urban growth, their African counterparts have opted to make the budgeting process more participatory.

That Johannesburg has made it somewhat successful sparks some hope that urban governance can improve for cities like Harare if and when there is a combination of favorable global and national circumstances.

NOTES

1. I would like to thank the Center for Urban and Global Studies at Trinity College for hosting me as a visiting scholar during 2008–09 when I started this research and writing. Xiangming Chen provided extensive comments and editing that has helped improve this chapter. Unless otherwise indicated by the citied sources, most of the qualitative information on the participatory budgeting of the two cities was obtained from extensive field work.
2. South Africa ranked first on a scale of democracy among eighteen African countries, whereas Zimbabwe ranked almost at the bottom (Devarajan, Khemani, and Walton 2011: Table 1).

REFERENCES

Chaeruka, Joel, and Peter Sigauke (2007). "Practitioners' Reflections on Participatory Budgeting in Harare, Mutoko and Marondera Workshops/Meetings and Experiences." *Local Governance and Development Journal* 1 (2): 5–28.

City of Johannesburg Report (2006). "Reflecting on a Solid Foundation: Building Developmental Local Government." Johannesburg.

Devarajan, Shantayanan, Stuti Khemani, and Michael Walton (2011). "Civil Society, Public Action, and Accountability in Africa." Washington, DC: The World Bank.

Fjeldstad, Odd-Helge (2004). "What's Trust Got to Do with It? Non-Payment of Service Charges in Local Authorities in South Africa." *Journal of Modern African Studies* 42 (4): 539–562.

Fjeldstad, Odd-Helge (2006). "To Pay, or not to Pay? Citizens' Views on Taxation by Local Authorities in Tanzania." Chr Michelsen Institute (CMI) Working Paper No. 8.

Human Rights Watch (2007). "The Human Rights Impact of Local Government Corruption and Mismanagement in Rivers State, Nigeria," 19 (2A).

Jameson, Boex, and Jorge Martinez-Vazquez (2003). "Local Government Reform in Tanzania:

Considerations for the Development of a System of Formula-Based Grants." *Andrew Young School of Policy Studies, Working paper 0305*, Atlanta.

Kamete, Amin Y. (2006). "More Than Urban Local Governance? Warring over Zimbabwe's Fading Cities." *African Renaissance* 3 (2): 34–46.

Langa, Bheki, and Afeikhena Jerome (2004). "Participatory Budgeting in South Africa." Working Paper Series, No. 1. Dakar: SISERA.

Marquetti, A.(2007), "The Characteristics of Brazilian Cities with Participatory Budgeting" <www.fee.tche.br/sitefee/download/jornadas/2/e/2-01.pdf, accessed on 9/14/08>.

Matovu, George, and Takawira Mumvuma (2008). "Participatory Budgeting in Africa: A Training Companion." Nairobi: UN-Habitat.

Mhlahlo, Rwadzi Samson (2007). "Assessment of Urban Governance in Zimbabwe, Case of the City of Gweru." *Eastern African Social Science Research Review* 23 (1): 107–128.

Nleya, Fanuel (2011). "Residents Praise City for Improved Service." *Newsday*, July 25.

Nyalunga, Dumisani (2007). "The Revitalization of Local Government in South Africa." *International NGO Journal* 1 (2): 15–20.

Putnam, Robert D. (1993). "The Prosperous Community: Social Capital and Public Life." *The American Prospect* 4 (13): 35–42.

Reinikka R. and Svensson, J. (2004) "Local Capture: Evidence from a Central Government transfer program in Uganda." *The Quarterly Journal of Economics*, pp 679–705 (2004).

Schugurensky, Daniel (2004). "Participatory Budgeting: A Tool for Democratizing Democracy." Talk given at the symposium "Some Assembly Required: Participatory Budgeting in Canada and Abroad," April 29.

Shah, Anwar, ed. (2007). *Participatory Budgeting.* Washington, DC: The World Bank.

Sintomer, Yves, Carsten Herzberg, Giovanni Allegretti, and Anja Röcke (2010). Learning from the South: Participatory Budgeting Worldwide—an Invitation to Global Cooperation, *InWEnt gGmbH—Capacity Building International*: Bonn.

Sten, Richard E., and Mohamed Halfani (2001). "The Cities of Sub-Saharan Africa: From Dependency to Marginality." Pp. 466–488 in *the Handbook of Urban Studies,* edited by Ronan Paddison. Glasgow: Sage Publications.

Toriro, P. (n.d.). "Africa Good Governance Programme on the Airwaves." Municipal Development Partnership for Eastern and Southern Africa, Harare.

Transparency International Zimbabwe (2009). (Press release) "Corruption at Harare

City Council," Harare: March 8.

UNEP (2002). *State of the Environment and Policy Retrospective, 1972–2002.* New York.

Wampler, Brian (2007). *Participatory Budgeting in Brazil, Contestation, Cooperation, and Accountability.* University Park, PA: University of Pennsylvania Press.

Wunsch, J.S (1998). "Decentralisation, Local Governance and the Democratic Transition in Southern: A Comparative Analysis", *African Studies Quarterly*, 2(1): 19–45

Yusuf, Fatima (2004). "Community Participation in the Municipal Budget Process: Two South African Case Studies." *Foundation for Contemporary Research (FCR) and Good Governance, Learning Network (GGLN).* South Africa: Paper presented at the LogoLink, International Workshop on Resources, Citizen Engagement and Democratic Local Governance (Porto Alegre, Brazil, December 6–9).

Yvette, Geyer (2007). *How Local Government Works.* Pretoria: IDASA.

Zimbabwe Gazette Daily News (2008). "CHRA Says No to Commission Proposals." Monday, December 17.

Part III

The Contested Urban Arena

Identity and Exclusion in Secondary Cities

9 Globalization and the Construction of Identity in Two New Southeast Asian Capitals

Putrajaya and Dompak

Sarah Moser

Over the past several decades, Southeast Asia has experienced rapid urbanization on a massive scale. With just 14 percent of Southeast Asians living in urban areas after World War Two, over 40 percent of the approximately 600 million Southeast Asians now live in cities, a percentage that is expected to exceed 55 percent by 2025 (United Nations 1995). Whereas Southeast Asia's urban growth since WWII has largely occurred in primate cities (King 2008), over the past two decades, urban growth has also taken the form of new cities built from a *tabula rasa*. New cities have been constructed primarily to nurture the manufacturing industry (Batam industrial estates in Indonesia, Shah Alam in Malaysia) and to encourage high-tech ambitions (Malaysia's Cyberjaya and the Multimedia Super Corridor, Can Tho City in Vietnam).

In recent years, cities are also being constructed as new seats of power to replace former colonial capitals. Bolstered by oil money, Malaysia and Indonesia[1] have prioritized the creation of two ostentatious new master-planned cities built on a grand scale: Putrajaya, Malaysia's new federal administrative capital, and Dompak, the new capital of Indonesia's Riau Islands, an archipelagic province formed in 2004. Started in 1995, Putrajaya is built on former oil palm and rubber plantations and is designed for 350,000 residents, primarily civil servants and their families. Dompak, the new capital of Riau Islands Province, was begun in 2007 and is at a much earlier stage of construction. Built on a 925 hectare (3.6 square miles) island, Dompak is located on the southern side of the current provincial capital, Tanjung Pinang, to which it will be connected by two bridges. Both cities replace the administrative functions of a vibrant and ethnically diverse seat of power established during the colonial era and represent broader development trends in Muslim-dominated Southeast Asia that link religion and national identity with the city and globalization. A study of the identities of Putrajaya and Dompak offers important insights into how local interpretations of globalization and the 'global city' have affected national development agendas and urban form in Malaysia and Indonesia (see Zukin, this volume, on a more general discussion on aspirations for global-city status).

While Southeast Asia's urban growth has not received as much attention as other regions of Asia, particularly China, there is a growing body of literature that examines varying dimensions of Southeast Asia's urban transformation. Scholars investigating Southeast Asian urbanization have directed their attention primarily towards the emergence of megacities and mega-urban regions (Bunnell 2002b; Bunnell 2004; Bunnell, Barter, and Morshidi 2002; Bunnell and Coe 2005; McGee and Robinson 1995; Silver 2008), transnational economic development (Bunnell, Muzaini, and Sidaway 2006; Grundy-Warr 2002; Grundy-Warr, Peachey, and Perry 1999; MacLeod and McGee 1996; Sparkle et al 2004; Yuan 1995), and 'creative clusters' and the knowledge economy (Kong 2009; Kong and O'Conner 2009; Wong and Bunnell 2006; Wong, Choi, and Millar 2006). In recent years there is a growing body of literature that examines the role of cities in the construction of identity in Southeast Asia (Chang 2000; Colombijn 1998; Colombijn 2003; King 2007; King 2008; Kong 2007; Kong and Yeoh 2003; Lim 2004; Moser 2007; Moser 2010b; Moser forthcoming). However, despite the exceptionally rapid growth and urban development of the Malay world[2] over the past decade, new Malay cities, including Putrajaya and Dompak, have received comparatively little scholarly attention. Recent research on the Riau Islands has focused primarily on transnational flows of labor and capital and the social and spatial implications of the Indonesia-Malaysia-Singapore Growth Triangle (Debrah, McGovern, and Budwar 2000; Grundy-Warr, Peachey, and Perry 1999; MacLeod and McGee 1996; Naidu 1998; Parsonage 1992; Sparke et al. 2004; Yuan 1995). While recent research on Malaysian urban growth has explored several recent megaprojects, less attention has been paid to the interplay between urban-centric national development agendas, nation-building efforts and globalization.

I examine Putrajaya and Dompak in order to investigate several issues. I seek to reveal trends in Southeast Asian symbolic national urbanism and offer a counternarrative to the literature focused on transnational economic policies that supposedly erode the authority of the nation. Whereas the significance of nation-states has been characterized as in decline (Ohmae 1992, 1995, 2001), I argue that despite an increasingly 'borderless' and 'global' world, the authority of the nation in Southeast Asia has in fact grown stronger in many ways as the ruling elite have devised new strategies to maintain and materialize the 'imagined community' of the nation (Anderson 1991 [1983]; Moser 2008, 2010a). In this chapter, the creation of new cities in the Malay world is placed in the broader context of globalization. While the Malay world's urban development does not compare with Dubai, Tokyo or Singapore in terms of global connections, size or economic importance, I demonstrate that small and secondary new cities such as Putrajaya and Dompak are intimately bound up in global processes and can be understood as a result of locally interpreted versions of globalization and the 'global-cities' discourse, and can in turn exert an international influence (see Kanna's comparison of Dubai and Singapore in this volume).

My article also draws attention to an emerging transnational sense of identity, specifically a recuperated sense of 'Malayness' manifested in Putrajaya and Dompak's generic Middle Eastern architectural idiom.

I begin with a brief overview of urbanism in the Malay world, providing a historical context for current urban developments. Second, I examine the role of 'serial seduction,' a process in which particular cities have a 'seductive' effect that inspires widespread emulation (Zukin, Masiya, this volume, on similar discussions). Third, I examine how globalization, local cultural politics, as well as local interpretations of 'global' have played a powerful role in shaping Putrajaya and Dompak, both discursively and materially. Finally, I examine how Putrajaya and Dompak, while not 'global cities' in the conventional sense, are intimately linked to global Islam.

URBANISM IN THE MALAY WORLD

Dunia Melayu, or the 'Malay world,' broadly consists of Malay-speaking Muslim areas of Southeast Asia, roughly considered to include the Malay Peninsula, coastal Borneo and Sulawesi, parts of Sumatra, Riau Islands Province and Singapore.[3] 'Malayness' is a complex and flexible identity that is based on religion (Islam), language and culture rather than on ethnicity (Barnard 2004) and, as such, can be interpreted visually in a variety of ways. The Malay world was historically linked by sea, including the Straits of Melaka, the channel between Sumatra and the Malay Peninsula that has served as a highway for maritime traders for nearly two millennia (Miksic and Low Mei Gek 2004; Reid 1993).

It is essential to contextualize the creation of Putrajaya and Dompak in the region's colonial history and postcolonial legacy as well as the dynamic between Singapore and its Malay neighbors. The Dutch and British, the colonial powers that controlled the Malay world in the nineteenth century, divided up what are now Malaysia, Singapore and Indonesia,[4] with the Dutch controlling present-day Indonesia and the British taking control of contemporary Malaysia and Singapore. The colonial powers established or greatly expanded a number of strategic trading ports and bureaucratic centers from which to administer their colonies, including Jakarta, Surabaya, Singapore, Kuala Lumpur, Georgetown, and Tanjung Pinang.

Starting in the mid-nineteenth century, the British used Kuala Lumpur as the bureaucratic base for their colony of Malaya due to its proximity to tin-mining operations, rubber plantations, and to the deep-sea port of Klang, through which materials were loaded onto ships for export. While Jakarta (then called Batavia) was the primary city of the Dutch colonial empire, numerous regional cities were developed over the course of three centuries of colonial rule to accommodate the growing colonial bureaucracy and ever-expanding maritime trade. The Dutch forcibly annexed the Riau Islands in 1911 and used the small trade entrepôt of Tanjung Pinang[5] as the political base from which to govern the Riau Archipelago. Tanjung Pinang was a strategic location for sea trade and easily protected, but more symbolically, it was the seat of the Riau sultanate, which fled to Singapore rather than be subservient to the Dutch colonial authorities.

Long-standing trade and family linkages across what are now Indonesia, Singapore and Malaysia were disrupted when colonial and, later, national boundaries made it more difficult for large numbers of people to move about the region freely following economic opportunities as they had for centuries. Postindependence nation-building in Indonesia, Malaysia and Singapore discouraged transnational Malay identity, and tensions between the three countries[6] served to fragment the region. However, a resurgent Malay nationalist movement reconstructed the notion of a transnational pan-Malay world (Kahn 2006), an identity that is used strategically by the Riau Islands and Malaysia, as will be seen in the context of Dompak and Putrajaya.

The Dutch and British colonial governments left a lasting impact on urban areas in Malaysia and Indonesia. Although vibrant precolonial cities existed (Reid 1993), contemporary urban Malaysia and the Riau Islands did not evolve from indigenous settlements, but from colonial administrative centers and international trade activities of the colonial governments and immigrants (Evers and Korff 2000). Most visibly, the colonial powers constructed buildings for their bureaucracy, many of which remain prominent features of cities in the Malay world. The British introduced a neoclassical 'British Raj' style incorporating elements of Victorian, neo-Gothic, Moorish and Mogul architecture. Characterized by fantastical domes, arches and arcades, it was referred to in colonial times as the 'Mahometan' style (King 2008). The Dutch developed an equally unique, but less ostentatious, 'Indische' (Indies) style of architecture that combined contemporary European styles, such as Art Deco, with the steeply peaked tile roofs indigenous to much of the Indonesian archipelago (Nas 2007).

Urbanism in the Malay world can be characterized by its uneven geographic distribution of ethnic groups, largely due to colonial racial philosophy. Colonial powers believed that humans could be divided into a hierarchy of races that were physically and mentally suited to different occupations (Alatas 1977). These beliefs served as the basis for a range of policies that influenced occupations, where different ethnic groups lived, and had a profound impact on the urban environment. Various late-colonial policies encouraged Malays, who were believed to be inherently 'lazy' and ill-suited to mercantile activities, to live in rural areas (Alatas 1977). Because Chinese were believed to be inherently more advanced in business, in the 1800s Chinese nationals were encouraged by colonial administrations to pursue mercantile activities in urban regions, a pattern that persists to the present day.

The geographic division of ethnic groups into racial enclaves gave distinct form to different neighborhoods. Starting from colonial times, the Chinese have resided in what are now Indonesian, Malaysian and Singaporean built 'shophouses,' a distinctive architectural form unique to Southeast Asia, in which the ground floor of a row house is used for commercial purposes with the upper few stories for residential use. The upper stories hang over the sidewalk and are supported by pillars, thus creating a shaded walkway. Chinese shophouses created dense and ordered urban neighborhoods that contrasted with the more dispersed, tree-filled Malay urban spaces (King 2008). Whereas the Chinese used stone and concrete building techniques brought from their more temperate homeland, Malays used local timber and indigenous building techniques developed over the centuries in response to the tropical climate for their homes, businesses, mosques and palaces (see Figure 9.1). The result is two vastly different urban forms which make the ethnicity of the inhabitants immediately apparent to the observer. Upon independence from colonial rule (1945 for Indonesia, 1957 for Malaya), Chinese residents constituted a majority of urban residents and, thanks largely to colonial policies, they controlled a majority of the urban economy in Malaysia and the Riau Islands until recent decades with the introduction of policies and incentives meant to attract Malays to urban areas. However, as has been pointed out by Thompson (2007), such policies have in fact resulted in the further creation of ethnic enclaves rather than a blending and sharing of urban space. It is against this backdrop of racial tension and geographic divisions that new 'Malay' cities of Putrajaya and Dompak were conceived.

Figure 9.1 Ethnic and religious diversity in urbanism of the Malay world: Hindu temple; typical row of Chinese shophouses; Chinese district, Tanjung Pinang; typical traditional Malay wood home (source: Author).

Figure 9.2 Global identities in Dompak: Computer model of Stadion Olahraga Gedebage (a Stadium Venue for Sports, Arts and Culture and Conventions), model of Dompak's Grand Mosque (this photo is labelled 'Dompak Island: The Singapore & Putrajaya of Tanjungpinang').

Source: Kepulauan Riau Government website, with research support from several local government employees to the author.

Powerful local cultural assumptions dating from colonial times foster and normalize urban-centric development in the Malay world. There is a deep-seated sense among the ruling elite and the general population that being 'successful' is inextricably linked with being urban. Cities are powerful symbols of progress in the Malay world, where the rural is often associated with being backwards and archaic whereas the urban is associated with modernity, progress and development. For example, *kampung* means village or neighborhood in Malay, and the related word *kampungan* literally means 'village-like' and translates as 'backwards,' 'boorish' or 'unsophisticated.' In this way, the drive to create new cities in the Malay world is a reflection of local urban-centric conceptualizations of modernity.

In the case of Malaysia and the Riau Islands, the desire to urbanize is now being realized thanks to a lucrative oil industry and governments that are keen to propel themselves into more powerful regional roles. Both governments have developed ambitious development and modernization agendas evidenced by massive spending on ostentatious and high-profile megaprojects. As part of Malaysian Prime Minister Mahathir's (1981–2003) Vision 2020 to fully modernize and develop the country by the year 2020, a number of megaprojects have been constructed since the mid-1990s with undisclosed budgets and designed by foreign 'superstar' architects. Mahathir's megaprojects include the Petronas Towers by Cesar Pelli (until recently, the tallest building in the world), the Multimedia Super Corridor, the Kuala Lumpur International Airport by Kisho Kurokawa, and the high-tech city of Cyberjaya. Similarly, the Riau Islands have constructed several megaprojects including a vast series of industrial estates centered primarily on Batam Island located thirty minutes by ferry from Singapore, a bridge project to connect a chain of (largely uninhabited) islands south of Batam Island, a massive zone for elite tourism along the northern coast of Bintan Island and the new provincial capital of Dompak. Still in the planning stages are several universities, a massive bridge to connect Batam Island with Bintan Island, and a sports stadium (see Figure 9.2). Both the oil revenues and the price tags for these projects are shrouded in secrecy, enabling the states to carry out their lavish building agendas with little accountability or public input.

GLOBALIZATION AND THE SEDUCTIVE CITY

Putrajaya and Dompak highlight the importance of what Bunnell and Das (2010: 2) refer to as the 'serial seduction' of cities, a geographic approach concerned with the 'seduction of urban projections of various forms . . . in shaping urban policy and material realities.' This approach departs from other urban geography inquiries in two key ways. First, while much of the geographic scholarship takes bounded urban areas as the unit of analysis, examining the process of 'serial seduction' focuses on the transnational connections between places and the multicentered nature of new urban development. Second, examining current patterns of 'serial seduction' makes it clear that, counter to much of the urban studies literature, innovations and models of city development do not necessarily originate in the west (Bunnell and Das 2010; Robinson 2006). King (1996) points out the process of 'Manhattan transfer' that occurred in the Asia Pacific region in the 1990s, when city authorities sought to replicate the skyline of New York City. More recently, however, completely different patterns of urban policy reproduction that look to other non-Western models are being created in the Malay world (see Zukin, this volume, on cities replicating cultural strategies as a sort of serial seduction).

The process of 'serial seduction' provides a useful conceptual handle with which to examine the decision to create Putrajaya and Dompak as well as their urban design and architectural idioms. Several key strands of urban replication can be observed in new cities. First, is the use of neighboring Singapore as a template. Important in the minds of the local elite in Malaysia and the Riau Islands, Singapore is a source of inspiration for ambitious government officials in the region who hope to replicate Singapore's economic success. Both Malaysia and the Riau Islands are, in their own ways, adopting Singapore's mantra of 'From Third World to First World,' a philosophy that encourages neoliberal development on a massive scale, ideally allowing the societies to 'leapfrog' from small-scale trading and manufacturing into a new knowledge-based economy. Among ruling elites, local bureaucrats and designers in the Riau Islands and parts of Malaysia, Singapore's unique micromanaged and ultraclean (even sterile) aesthetic is imagined as being a feature of 'global cities' and has been reproduced in many new developments. Singapore's clean tile sidewalks, the clusters of pastel apartment blocks organized into 'estates,' and the long roster of rules reproduced in many new developments in Malaysia and the Riau Islands all reflect Singapore's role as regional urban role model. In contrast to the labyrinthine colonial-era urban fabric of Kuala Lumpur and Tanjung Pinang, it is Singapore's wide avenues and expansive highway system that have been a template for recent development in Malaysia and Indonesia.

Second, colonial-era Garden City ideals, which have been circulating widely in Southeast Asia for nearly a century, have been replicated in Putrajaya and Dompak, particularly in the prioritization of elaborate parks and gardens. Botanical gardens, a colonial invention of which Singapore has the most spectacular example in the region, are experiencing a renaissance in state-planned cities in the Malay world. Putrajaya and Dompak both have expansive Taman Botani (Botanical Gardens) that are clearly understood by local officials as a symbol of prestige and 'global' sophistication.

Third, as the most ostentatious capital in Southeast Asia to date and the benchmark in the region for what a grand Southeast Asian capital can be, Putrajaya has become a model of a progressive Muslim city both for the region and for much of the Muslim world. Putrajaya has inspired or 'seduced' government officials within the Malay world and the Muslim world more broadly to create their own competing version of a Malay/Muslim city. Furthermore, while few 'global-cities' scholars have heard of Putrajaya, the city is becoming internationally known and highly influential among postcolonial nations in Southeast Asia and in much of the Muslim world that aspire to be 'global' yet retain a sense of tradition and religious values (Moser 2010b). Although it has been dismissed as a folly by the press and academia, Putrajaya is an important new symbol in the region and has set a new standard for Malay(sian) urbanism, resulting in many 'little Putrajayas' springing up in Malaysia and Indonesia, including Dompak (Tajuddin 2005). Putrajaya is widely seen in the Muslim world as a model progressive 'Islamic' city that is grounded in religious values that are expressed in a recognizably Islamic idiom, while nurturing modernity and high-tech ambitions. In recent years, government officials, planners, architects and students from Muslim regions of Africa and Central Asia have traveled to Putrajaya to view the city firsthand and several recently built seats of power have clearly been influenced by Putrajaya, including Kazakhstan's new capital of Astana (see Moser forthcoming).

Finally, city officials in charge of creating Putrajaya and Dompak have been seduced by the visioning process itself, which places more emphasis on image, identity and 'world-class' features than on the inevitable sociospatial fragmentation that such grandiose state plans create. It is significant, however, that it is not the traditional 'global cities' (e.g., London, New York, Tokyo) that are being looked to for 'global' inspiration, but new regional urban models. Cities that do not rank among the 'global cities' are in fact more influential for the new cities in the Malay world than the established financial centers that are the focus of academic research.

REFRACTING THE GLOBAL THROUGH THE LOCAL: NEW CITIES OF GLOBAL ISLAM

As Sharon Zukin argues in this book, in many ways globalization is having a homogenizing affect on the appearance of cities as more office space and ostentatious architectural trophies are created in an effort to attract corporate and elite international tenants, as cities attract the same luxury shops, and as similar leisure landscapes are built. Cities are increasingly understood as key players driving economic change (Castells 1996), and a growing number of governments consider cities as powerful tools to attract headquarters of regional branches, international investors, tourists, and foreign talent (Bunnell 2002a). As a result, cities are progressively being viewed as products that need to be marketed and promoted to the world through urban spectacles, megaprojects and record-breaking architectural feats in an attempt to both set themselves apart from other cities while conforming to international expectations about what a 'global city' should look like (Elsheshtawy 2010). In this context of one-upmanship, records for the world's tallest building have been held by three countries in Asia in the past decade, architectural megaprojects are increasingly lavish and spectacular, and gigantic Ferris wheels mark the skyline of a growing number of (aspiring) 'global cities.'

Figure 9.3 Putrajaya as Islamic capital: Putra Mosque, *Komplex Kehakiman* (Palace of Justice), prime minister's office (source: Author).

While I do not dispute Zukin's point about increasing homogenization among 'global' cities, I suggest that there is an emerging network of smaller, second- or third-tier cities in the Muslim world that have no possibility of attracting regional corporate head offices or luxury shops yet have been influenced by local interpretations of globalization and the 'global-cities' discourse. The result is in the emergence of small new cities, including

Figure 9.4 Dompak as Islamic capital—provincial government office complex, Islamic cultural center (source: Author).

Putrajaya and Dompak, which display unique regional identities and often unexpected versions of 'global' features. These new cities are influenced by the increase in communication, transportation, and economic links that characterize globalization, yet their appearance has not become homogenized. Rather, they are also a response to the homogeneity of 'placelessness' produced by globalization and are a way to convey local cultural values. Shim (2006: 27) and others (Featherstone 1993; Robertson 1995) argue that paradoxically, globalization encourages local peoples to discover the 'local' that they have neglected or forgotten in their drive towards Western-imposed modernization during the past decades.

Over the past decade, this has resulted in the widespread 'invention of tradition' (Hobsbawm and Ranger 1983) manifested in the creation of a range of pseudotraditional and cultural projects, including ethnicized airports featuring temples, ancient-looking statuary, and faux night markets; the designation, enhancement and exoticization of urban cultural districts; the use of 'traditional' architecture in modern buildings; and many other such projects.

In urban areas of the Malay world, particularly in new cities such as Putrajaya and Dompak, many recent state projects have adopted a generic Middle Eastern architectural idiom,[7] transforming skylines into a sort of 'Arabian Nights' (Moser 2010b, Moser forthcoming). The domes, arches and minarets found on government buildings, bridges and shopping malls in Putrajaya are inspired by an imagined Middle East and are taken from such diverse sources as Ottoman, Savafid, Central Asian, Iraqi, Persian, Moorish Spanish and Mughali architecture, resulting in an eclectic hodgepodge of recognizably 'Islamic' architecture and urbanism (see Figure 9.3). While Dompak is at an earlier stage than Putrajaya, the architectural models and buildings in progress reveal a similar adoption of Middle Eastern forms[8] (see Figure 9.4). Putrajaya and Dompak are important examples of the attempt to create an 'authentic' and 'exotic' cultural spectacle for the purposes of city branding, aimed at both an international tourism audience and for nation-building purposes.

Putrajaya and Dompak are fascinating local manifestations of the rise of global pan-Islam. While the 'fantasy Middle Eastern' style has received some criticism from the media and scholars, it is currently the most widely used style for new urban areas in the Malay world. Scrupulous attention is paid to the selection of domes and arches and 'Islamic' details, such as orienting new cities to face Mecca, a device that has never been used in the history of Islam. This Islamic turn should be understood in the context of growing connections between the Malay world and the Middle East through 'Islamic' banking, development partnerships, and growing political ties. I suggest that, through adopting a generic Middle Eastern idiom, Putrajaya and Dompak are positioning themselves to attract the attention and, more importantly, the business of 'global' Muslims (i.e., the wealthy Arab elite) through tourism and investment.

Whereas Putrajaya and Dompak are designed to assert a Malay/Muslim identity over a general sense of disenfranchisement at the hands of colonial powers and, more importantly, local Chinese populations, they demonstrate a global awareness of a broad, transnational Islam. As numerous scholars have pointed out, the global-cities literature ranks and classifies cities on purely economic criteria (for example, the number of banks, law firms, and headquarters for transnational corporations based in a city) (Elsheshtawy 2010) in a way that ultimately serves to privilege primarily European and American cities (Robinson 2002). Elsheshtawy (2008) argues that the narrow economic focus of the global-cities discourse excludes cities such as Jerusalem, Medina, and Mecca that are surely global in the sense that they serve as religious centers which attract millions of international pilgrims annually. However, since they do not host headquarters for multinational corporations, they are not considered 'global' cities.

Following Elsheshtawy, I suggest that the impact global Islam has had on cities in the Muslim world cannot be underestimated (also see Moser forthcoming). Significantly, like transnational corporations, global pan-Islam is increasingly focused on the scale of the city and invested in image and identity to represent 'Islamic' identity and transnational solidarity. Furthermore, the similarities between both Putrajaya and Dompak and the cities they are designed to replace are striking and fit into broader patterns in the Muslim world. Significantly, like many cities in the Muslim world, Kuala Lumpur and Tanjung Pinang are characterized by their ethnically and religiously diverse populations and are dotted with mosques, temples and churches with Chinese constituting the largest group. Putrajaya and Dompak are designed to send a powerful message to the non-Malay populations and to assert Muslim dominance over a diverse citizenry.

Against this backdrop of phenomenal change, both Malaysia and the Riau Islands have attempted to balance massive development with maintaining aspects of traditional Malay identity and values. A key part of this effort to maintain 'Malayness' can be seen in the creation of new capital cities that make strong statements about cultural and religious values. Far from the placelessness associated with urban development and globalization, these new cities are developing a unique (if constructed) regional character of 'Malayness' in order to replace existing seats of power created (or vastly expanded) during colonial occupation.

THE MANY MEANINGS OF 'GLOBAL'

The concepts of globalization and 'global cities' are highly contested among academics and city officials in globalizing nations that have interpreted the concepts in a variety of unexpected ways. Whereas many scholars of 'global cities' and globalization have tended to focus on dominant cities of the world urban hierarchy as key nodes in transnational economic, social and technological networks (Sassen 1991; Yeoh 1999), others suggest that the concept of a 'global city' has been used uncritically and that the term needs to be unpacked in other cultural contexts. Elsheshtawy (2010) argues that little attention has been paid to the phenomenon of cultural globalization and the traveling of culture, religion and ethnicity through the design of cities, and cities ranked by "cultural globalization" (Benton-Short, Price, and Friedman 2005). Peter Smith underscores the constructedness of the term 'global cities' when he argues that

> there is no solid object known as the 'global city' appropriate for grounding urban research, only an endless interplay of differently articulated networks, practices, and power relations—the global cities discourse constitutes an effort to define the global city as an objective reality operating outside the social construction of meaning. (2001: 378)

Similarly, other scholars have argued that the concept of 'global cities' obscures other important aspects of city life and suggest that 'ordinary' cities can be used profitably to examine transnational connections (Amin and Graham 1997; Robinson 2002). As Robinson (2002) argues, many cities will never be 'command and control' centers yet have been influenced by 'global cities' in various, often unanticipated, ways.

The neoliberal economic policies adopted by the ruling elite to create 'global' cities have transformed cities around the world and, as Sharon Zukin (2009) argues, tend to make cities look the same. However, it is important to note that there are 'slippages' between the official 'global-cities' discourse and the way in which local officials creating new cities reinterpret what makes a city 'global.' There is recently greater recognition that local agents shape, reinterpret, and interact with global processes in various ways (Marcuse and van Kempen 2000).

For many cities aspiring to 'globalize,' the 'global-cities' discourse has produced a sort of checklist of features that are perceived to be needed in order to be truly 'global.' This checklist includes becoming an international air hub, hosting global or regional corporate headquarters and banks, and developing international-quality facilities for business, culture and leisure. However, having interviewed government officials in Putrajaya and the Riau Islands,[9] it is apparent that the checklist contains rather different elements based on regional cultural values. Such elements are imagined as 'global' or 'international' features yet are not necessarily included in academic conceptualizations of the 'global city.' As mentioned, Singapore is the region's 'global city' and the benchmark to which many Southeast Asian cities aspire. However, the features that make Singapore a 'global city' to scholars (regional corporate headquarters, command-and-control center for transnational commerce, etc.) are not necessarily the features that Indonesian and Malaysian government bureaucrats prioritize.

In the case of Putrajaya and Dompak, many local elite interpret being 'global' through the construction of various urban design and architecture projects that generally evoke a sense of the 'international' and the foreign. In contrast to the labyrinthine streets of Kuala Lumpur and Tanjung Pinang, Putrajaya and Dompak prioritize wide roads and are each designed around grand oversized kilometers-longs ceremonial axes. As the head architect of Putrajaya explained, the designers began with an axis because 'capital cities have axes' (interview cited in King 2008: 150). In much of Southeast Asia, manicured landscaping also carries the connotation of prestige and wealth and both cities have adopted a highly controlled landscape aesthetic using decorative plants and topiary.

Similar to other globalizing cities, Dompak and Putrajaya adopt an architectural vocabulary of the spectacular in an attempt to overshadow neighboring cities and the former capitals they are designed to replace, but also to construct a distinct identity. However, as examined in the previous section, the urban spectacles take the shape of elaborate fantasy-style Islamic architecture produced by local designers, rather than in the form of the architectural trophies designed by international 'starchitects' commissioned in such 'global cities' as London and New York.

City officials in Putrajaya and Dompak also perceive that they are competing not with 'global cities' around the world but with nearby cities in the region. In fact, official material for Dompak refers to it rather unambiguously as 'The Singapore and Putrajaya of Tanjung Pinang' and 'The Miniature of Singapore inside Tanjung Pinang'! This regional competition takes various forms: longer and wider ceremonial axes, larger and more ostentatious mosques, and ever-more-extravagant conference facilities. One government official with whom I spoke in Riau felt that a major stadium and a Sirkuit International, or International Circuit for motor sports events, would boost Dompak's profile and its 'international' (i.e., global) qualities. Motor sports are tremendously popular in the region, and both Malaysia and Singapore have hosted prestigious international events in recent years. Motor sports are so popular that during political campaigns, some candidates will have a portion of the city shut down while Formula One race cars drive a circuit around the city. Such events draw massive crowds, and it is hoped that the thrills provided will translate into votes.

In the case of Putrajaya and Dompak, these local interpretations of the 'global' highlight the fact that city officials have generally only traveled to cities within the Southeast Asian region. More importantly, these cities demonstrate how third-tier cities that will never be 'global' cities according to conventional economic definitions are still influenced both materially and discursively through globalization and the 'global-cities' discourse. Despite the hegemonic nature of globalization, small new cities such as Putrajaya and Dompak demonstrate the slippages and local reinterpretations of 'global' elements that are resulting in unique regional urban patterns.

THE FUTURE OF PUTRAJAYA AND DOMPAK

The nature of top-down state projects such as Putrajaya and Dompak allows insight into how the ruling elite respond to globalization, attempt to nurture a locally interpreted version of what it means to be 'global' and how efforts to be 'global' are intertwined with nation-building. Putrajaya and Dompak are unique in their distinct need to symbolize the dominance of Malays over Chinese. Beyond local politics, both cities are striving to be 'international quality,' however, in a way that is a local interpretation. In this way, it is helpful to ground research on globalization through recognition that the 'global' is locally produced and reified through city design. Even if the intention is to directly reproduce aspects of the 'global city,' slippages are inevitable as local agents modify and reinterpret the meaning.

The highly urban-centric development approach of the Malaysian and Indonesian states is in itself strongly influenced by globalization and neoliberal development ideology. The attempt to build themselves into significance through constructing ostentatious new cities responds to the current global climate of producing grandiose urban spectacles and the perception of competing with other cities. Significantly, Putrajaya and Dompak imagine themselves as competing with other grand capitals around the world, including Singapore. However, in response to the placelessness and homogeneity produced by globalization, the cities have adopted a distinct 'local' architectural idiom of a generic and imagined Middle East to represent national identity.

While influenced by globalization and the 'global-cities' discourse, Dompak and Putrajaya are local interpretations of what constitutes 'global' that are patterned not after the conventional global cities of London, Paris, and New York but on urban examples imagined or found in the region. The goals of these cities are also not 'global' in the conventional sense. Government officials of Dompak and Putrajaya and the designers they employ seek to align themselves with global Islam.

In the present era of globalization and urban-centric, neoliberal development, such state-driven megaprojects as Putrajaya and Dompak are likely to continue to be prioritized in Southeast Asia. Around the world, megaprojects are seen as a way to jump-start economies, to leapfrog from Third World to First, and are seen as necessary in competing for foreign direct investment. In Southeast Asia, an increasing amount of stock is placed in having a showpiece capital to display both to the citizenry and to the international community that functions as a metonym for the state and its potential to balance modernization with maintaining traditional values and identity. In this way, global aspirations are closely linked with nation-building and are materialized through ever more ostentatious master-planned cities.

NOTES

1. Myanmar's military regime unveiled its new capital city of Naypyidaw in 2008, replacing the colonial capital of Yangon (formerly Rangoon). Little is known about this new city, although Donald Seekins has written an insightful short paper on Naypyidaw (Seekins 2009).
2. The Malay world, or *Dunia Melayu*, is a concept developed by Malay nationalists in the 1920s in an attempt to consolidate Malay peoples against the colonial Dutch and British and the Chinese (Kahn 2006).
3. Singapore is about 80 percent Chinese, 13 percent Malay and 7 percent various Indian groups yet is still considered by many Malays to be part of *Dunia Melayu*. This is part of the contemporary imaginings of historical Malays as a homogenous, united entity, when in fact as late as the nineteenth century Malays 'were concerned primarily with their differences rather than their shared features' (Milner 1995: 14).
4. The Anglo-Dutch Treaty of 1824 formally divided the British and Dutch territories in Southeast Asia as a means to prevent further wars between the two countries. The boundary separating the British and Dutch colonies eventually became the border separating Malaysia and Indonesia (Webster 1998).
5. Now a city of 150,000, Tanjung Pinang is still focused on regional maritime trade and is the largest and most important city in the Riau Islands.
6. Singapore was forced out of Malaya in 1965 and became an independent city-state, after which Malaya slightly changed its colonial name to Malaysia. From 1963 to 1966, long-term hostility between Indonesia and the British colonial authorities culminated a military confrontation, commonly referred to as *Konfrontasi*, between Indonesia and Malaya. At the end of WWII, the British, trying to reclaim the Indies on behalf of the Dutch, attempted to purge land of Indonesian nationalist influences through a series of land and air provocations in Kalimantan, Sumatra and Riau. In response, Indonesian president Sukarno ordered navy ships into Malayan waters (Poulgrain 1998).
7. I examine the use of generic Middle Eastern design in Putrajaya through the rubric of self-Orientalization in a forthcoming paper in *Urban Studies*: 'Circulating Visions of 'High Islam': The Adoption of Fantasy Middle Eastern Architecture in Constructing Malaysian National Identity.'
8. This can also be seen in the first buildings in the new administrative suburb of Senggarang, built to shift city government functions away from Tanjung Pinang, the current capital of Riau Islands Province.
9. I have conducted field research in Putrajaya in 2007 and 2008 and carried out intensive ethnographic fieldwork in the Riau Islands in 2006–08.

REFERENCES

Alatas, Syed Hussein (1977). *The Myth of the Lazy Native: A Study of the Image of the Malays, Filipinos and Javanese from the 16th to the 20th Century and Its Function in the Ideology of Colonial Capitalism*. London: Frank Cass.

Amin, Ash, and Stephen Graham (1997). "The Ordinary City." *Transactions of the Institute of British Geographers* 22 (4): 411–429.

Anderson, Benedict (1991) [1983]. *Imagined Communities: Reflections on the Origin and Spread of Nationalism*. London/New York: Verso.

Barnard, Timothy P., ed. (2004). *Contesting Malayness: Malay Identity across Boundaries*. Singapore: Singapore University Press.

Benton-Short, Lisa, Marie D. Price, and Samantha Friedman (2005). "Globalization from Below: The Ranking of Global Immigrant Cities." *International Journal for Urban and Regional Research* 29 (4): 945–959.

Bunnell, Tim (2002a). "Cities for Nations? Examining the City-Nation-State Relation in Information Age Malaysia." *International Journal for Urban and Regional Research* 26 (2): 284–298.

Bunnell, Tim (2002b). "Multimedia Utopia? A Geographical Critique of High-Tech Development in Malaysia's Multimedia Super Corridor." *Antipode* 34 (2): 265–295.

Bunnell, Tim (2004). *Malaysia, Modernity, and the Multimedia Super Corridor: A Critical Geography of Intelligent Landscapes*. London: Routledge.

Bunnell, Tim, Paul A. Barter, and Sirat Morshidi (2002). "Kuala Lumpur Metropolitan Area: A Globalizing City-Region." *Cities* 19 (5): 357–370.

Bunnell, Tim, and Neil M. Coe (2005). "Re-Fragmenting the 'Political': Globalization, Governmentality and Malaysia's Multimedia Super Corridor." *Political Geography* 24 (7): 831–849.

Bunnell, Tim, and Diganta Das (2010). "Urban Pulse—A Geography of Serial Seduction: Urban Policy Transfer from Kuala Lumpur to Hyderabad." *Urban Geography* 31 (3): 1–7.

Bunnell, Tim, Hamzah Muzaini, and James D. Sidaway (2006). "Global City Frontiers: Singapore's Hinterland and the Contested Socio-Political Geographies of Bintan, Indonesia." *International Journal for Urban and Regional Research* 30 (1): 3–22.

Castells, Manuel (1996). *The Rise of the Network Society*. London: Blackwell.

Chang, T. C. (2000). "Renaissance Revisited: Singapore as a 'Global City for the Arts.' " *International Journal for Urban and Regional Research* 24 (4): 818–831.

Colombijn, Freek (1998). "A Sheep in Wolf's Clothing." *International Journal for Urban and Regional Research* 22 (4): 565–581.

Colombijn, Freek (2003). When There Is Nothing to Imagine: Nationalism in Riau." Pp. 333–370 in *Framing Indonesian Realities; Essays in Symbolic Anthropology in Honour of Reimar Schefold*, edited by Peter J. M. Nas, Gerard A. Persoon, and Rivke Jaffe. Leiden: KITLV Press.

Debrah, Yaw A., Ian McGovern, and Pawan Budwar (2000). "Complementarity or Competition: The Development of Human Resources in a South-East Asian Growth Triangle: Indonesia, Malaysia and Singapore." *The International Journal of Human Resource Management* 11 (2): 314–335.

Elsheshtawy, Yassar (2010). *Dubai: Behind an Urban Spectacle*. London and New York: Routledge.

Elsheshtawy, Yassar, ed. (2008). *The Evolving Arab City: Tradition, Modernity & Urban Development*. New York and London: Routledge.

Evers, Hans-Dieter, and Rüdiger Korff (2000). *Southeast Asian Urbanism: The Meaning and Power of Social Space*. Singapore: Institute of Southeast Asian Studies.

Featherstone, Mike (1993). "Global and local cultures." Pp. 169–187 in *Mapping the Futures: Local Cultures, Global Change*, edited by John Bird, Barry Curtis, Tim Putnam, and Lisa Tickner. London: Routledge.

Grundy-Warr, Carl (2002). "Cross-Border Regionalism through a "South-East Asian" Looking Glass." *Space and Polity* 6 (2): 440–460.

Grundy-Warr, Carl, K. Peachey, and Martin Perry (1999). "Fragmented Integration in the Singapore-Indonesian Border Zone: Southeast Asia's 'Growth Triangle' against the Global Economy." *International Journal of Urban and Regional Research* 23 (2): 304–328.

Hobsbawm, David, and Terebce Ranger, eds. (1983). *The Invention of Tradition.* Cambridge: Cambridge University Press.
Kahn, Joel S. (2006). *Other Malays: Nationalism and Cosmopolitanism in the Modern Malay World.* Copenhagen: NIAS Press.
King, Anthony D. (1996). "Worlds in the City: Manhattan Transfer and the Ascendance of Spectacular Space." *Planning Perspectives* 11: 97–114.
King, Ross (2007). "Re-Writing the City: Putrajaya as Representation." *Journal of Urban Design* 12 (1): 117–138.
King, Ross (2008). *Kuala Lumpur and Putrajaya: Negotiating Urban Space in Malaysia.* Copenhagen: NIAS Press.
Kong, Lily (2007). "Cultural Icons and Urban Development in Asia: Economic Imperative, National Identity, and Global City Status." *Political Geography* 26: 383–404.
Kong, Lily (2009). "Making Sustainable Creative/Cultural Space in Shanghai and Singapore." *Geographical Review* 91 (1): 1–22.
Kong, Lily, and Justin O'Conner, eds. (2009). *Creative Economies, Creative Cities: Asian-European Perspectives.* London: Springer.
Kong, Lily, and Brenda S. A. Yeoh (2003). *The Politics of Landscapes in Singapore: Constructions of 'Nation.'* Syracuse, NY: Syracuse University Press.
Lim, William S. W. (2004). *Architecture Art Identity in Singapore: Is There Life after Tabula Rasa?* Singapore: Asian Urban Lab.
MacLeod, Scott, and Terence G. McGee (1996). "The Singapore-Johor-Riau Growth Triangle: An Emerging Extended Metropolitan Region." Pp. 417–464 in *Emerging World Cities in Pacific Asia*, edited by Fu-chen Lo and Yue-man Yeung. Tokyo: United Nations University Press.
Marcuse, Peter, and Ronald van Kempen (2000). *Globalizing Cities: A New Spatial Order.* Oxford: Blackwell.
McGee, Terence G., and Ira M. Robinson, eds. (1995). *The Mega-Urban Regions of Southeast Asia.* Vancouver: UBC Press.
Miksic, John N., and Cheryl-Ann Low Mei Gek, eds. (2004). *Early Singapore 1300s–1819: Evidence in Maps, Text and Artefacts.* Singapore: Archipelago Press.
Milner, Anthony C. (1995). *The Invention of Politics in Colonial Malaya: Contesting Nationalism and the Expansion of the Public Sphere.* Cambridge: Cambridge University Press.
Moser, Sarah (2007). "Reclaiming the City: The Fourth Great Asian Streets Symposium." *Cities: The International Journal of Urban Policy and Planning* 24 (3): 242–244.
Moser, Sarah (2008). *Performing National Identity in Postcolonial Indonesia.* Department of Geography, National University of Singapore, Singapore.
Moser, Sarah (2010a). "Creating Citizens Through Play: The Role of Leisure in Indonesian Nation-building." *Social and Cultural Geography* 11 (1): 53–73.
Moser, Sarah (2010b). "Putrajaya: Malaysia's New Federal Administrative Capital." *Cities: The International Journal of Urban Policy and Planning* 27 (3): 1–13.
Moser, Sarah (forthcoming). "Circulating Visions of 'High Islam': The Adoption of Fantasy Middle Eastern Architecture in Constructing Malaysian National Identity." *Urban Studies.*
Moser, Sarah, ed. (forthcoming). *New Cities in the Muslim World.* London: Reaktion.
Naidu, G. (1998). "Johor-Singapore-Riau Growth Triangle: Progress and Prospects." Pp. 29–49 in *Growth Triangles in Asia: A New Approach to Regional Economic Cooperation* (2nd Ed.), edited by Myo Thant, Min Tang, and Hiroshi Kakazau. Oxford: Oxford University Press.

Nas, Peter J. M., ed. (2007). *The Past in the Present: Architecture in Indonesia.* Rotterdam: NAi Publishers.

Ohmae, Kenichi (1992). *Borderless World: Power and Strategy in the Global Marketplace.* London: HarperCollins.

Ohmae, Kenichi (1995). *The End of the Nation State: The Rise of Regional Economics.* New York: The Free Press.

Ohmae, Kenichi (2001). "How to Invite Prosperity from the Global Economy into a Region." In *Global City-Regions: Trends, Theory, Policy*, edited by Allen J. Scott. New York: Oxford University Press.

Parsonage, James (1992). "Southeast Asia's 'Growth Triangle': A Sub-Regional Response to a Global Transformation." *International Journal of Urban and Regional Research* 16: 42–63.

Poulgrain, Greg (1998). *The Genesis of Kontrontasi: Malaysia, Brunei and Indonesia, 1945–1965.* Bathurst, UK: Crawford House Publishing.

Reid, Anthony (1993). *Southeast Asia in the Age of Commerce 1450–1680, Volume Two: Expansion and Crisis.* New Haven and London: Yale University Press.

Robertson, Roland (1995). "Glocalization: Time-Space and Homogeneity-Heterogeneity." Pp. 25–44 in *Global Modernities*, edited by Mike Featherstone, Scott M. Lash, and Roland Robertson. London: Sage.

Robinson, Jennifer (2002). "Global and World Cities: A View from off the Map." *International Journal for Urban and Regional Research* 26 (3): 531–554.

Robinson, Jennifer (2006). *Ordinary Cities: Between Modernity and Development.* Abingdon, UK: Routledge.

Sassen, Saskia (1991). *The Global City: London, New York, Tokyo.* Princeton, NJ: Princeton University Press.

Seekins, Donald M. (2009). " 'Runaway Chickens' and Myanmar Identity: Relocating Burma's Capital." *City* 13 (1): 63–70.

Shim, Doobo (2006). "Hybridity and the Rise of Korean Popular Culture in Asia." *Media, Culture & Society* 28 (1): 25–44.

Silver, Christopher (2008). *Planning the Megacity: Jakarta in the Twentieth Century.* New York: Routledge.

Smith, Michael Peter. (2001). *Transational Urbanism: Locating Globalization.* Malden, MA: Blackwell.

Sparke, Michael, James D. Sidaway, Tim Bunnell, and Carl Grundy-Warr (2004). "Triangulating the Borderless World: Geographies of Power in the Indonesia-Malaysia-Singapore Growth Triangle." *Transactions of the Institute of British Geographers* 29: 485–498.

Tajuddin, Mohammad Haji Mohamad Rasdi (2005). *Malaysian Architecture: Crisis Within.* Kuala Lumpur: Utusan Publications & Distributors.

Thompson, Eric C. (2007). *Unsettling Absences: Urbanism in Rural Malaysia.* Singapore: National University of Singapore Press.

United Nations (1995). *World Urbanization Prospects: The 1994 Revision—Estimates and Projections of Urban and Rural Populations and of Urban Agglomerations.* New York: United Nations.

Webster, Anthony (1998). *Gentlemen Capitalists: British Imperialism in Southeast Asia, 1770–1890.* New York: Tauris Academic Studies.

Wong, Caroline Y. L., Chong Ju Choi, and Carla C. J. M. Millar (2006). "The Case of Singapore as a Knowledge-Based City." Pp. 87–96 in *Knowledge Cities: Approaches, Experiences and Perspectives*, edited by Francisco Javier Carrillo. Oxford: Butterworth-Heinemann.

Wong, Kai Wen, and Tim Bunnell (2006). "'New Economy' Discourse and Spaces in Singapore: A Case Study of One-North." *Environment and Planning A* 38 (1): 69–83.

Yeoh, Brenda. (1999). "Global/globalizing cities." *Progress in Human Geography.* 23 (4): 607–616.

Yuan, Lee Tsao (1995). "The Johor-Singapore-Riau Growth Triangle: The Effect of Economic Integration." Pp. 269–281 in *The Mega-Urban Regions of Southeast Asia*, edited by Terence G. McGee and Ira M. Robinson. Vancouver: UBC Press.

Zukin, Sharon (2009). "Destination Culture: How Globalization Makes All Cities Look the Same." *Center for Urban and Global Studies Working Paper* Series 1 (1): 1–26.

INTERNET REFERENCES

Pemerintah Provinsi Kepulauan Riau. [Government of Riau Islands Province] '*Pembangunan Pusat Pemerintahan di Pulau Dompak*' [Building a Government Center on Dompak Island] (accessed 3 January 2010) http://www.kepriprov.go.id/id/?option=com_content&task=view&id=247&Itemid=41.

Perbadanan Putrajaya. (accessed 24 November, 2009) http://www.ppj.gov.my/portal/page?_pageid=311,1&_dad=portal&_schema=PORTAL

Portal Nasional Republik Indonesia. [National Portal for the Republic of Indonesia] '*Pembangunan pusat pemerintahan Provinsi Kepri*' [Building a government center for Riau Islands Province] (accessed 24 November 2009) http://www.indonesia.go.id/id/index.php?option=com_content&task=view&id=4626&Itemid=832.

Putrajaya Holdings. http://www.pjh.com.my/ (accessed 24 November 2009).

10 Off Limits and Out of Bounds

Taxi Driver Perceptions of Dangerous People and Places in Kunming, China

Beth E. Notar

In *Consciousness and the Urban Experience*, David Harvey asks of Haussman's demolished and reconstructed Paris: "How did people view each other, represent themselves and others to themselves and others? How did they picture the contours of . . . society, comprehend their social position and the radical transformations then in progress?" (Harvey 1985: 180). This chapter poses a similar set of questions to the contemporary Chinese city. In the context of the phenomenal growth of the Chinese economy, the radical reshaping of urban space, and a massive influx of rural-to-urban migrants, how do Chinese urbanites imagine themselves and others moving in these new geographic, sociocultural and economic spaces? And what are the implications of these imaginings?

Specifically, this chapter examines the views of taxicab drivers in the southwestern Chinese city of Kunming—a growing metropolitan area of over six million persons—toward what they consider to be dangerous people and places: people they try to avoid letting ride in their cabs and places where they try to avoid driving. Whereas urban scholars have examined elite "techniques of exclusion" and fears of the city in the Americas (e.g., Caldeira 2001; Lowe 2001, 2003), we have paid less attention to the ways in which nonelites might participate in exclusionary discourses and practices. And whereas anthropologists and sociologists of China have examined the lives of rural-to-urban migrant workers as well as prejudices against them (e.g., Pun 2005; Solinger 1999a, b; Yan 2008; Zhang 2001a, b), we have not fully examined the specific contours of these prejudices by nonelites.

Here I am interested in how people negotiate transformed cityscapes and imagine themselves and others as they move through the city. James Donald has argued that the "*living* space of the city exists as projection and experience as much as it exists as bricks and mortar or concrete and steel" (Donald 1997: 182). Taking inspiration from Donald, I am listening to "narratives about cities" which "project a narrational space" onto the urban form. As Donald emphasizes, we should not expect these narratives to paint an "accurate portrait of the city" (Donald 1997). Instead, the narratives reveal the ways in which different groups and individuals perceive and experience the city. I will argue that if we are to understand the geographic imaginaries and conceptual and physical processes of boundary making in transformed cityscapes, we must pay attention not only to the "discourses of fear" and the "forms of exclusion and enclosure" practiced by elites (Caldeira 2001: 1); we must also understand the ways in which nonelites are actively involved in drawing boundaries and limiting the mobility of others.

Due to their intense physical mobility—Kunming taxi drivers move through the city nearly twelve hours a day, six days a week—I am particularly interested in how taxi drivers imagine and narrate a radically transformed city. As I myself attempted to move through the city in cabs, I was often caught in a traffic jam for thirty minutes to an hour.[1] And as I spent more and more time talking with cabdrivers as we were stuck in traffic, I became fascinated in what I call here their *discourses of avoidance*—where they said they tried to avoid driving and whom they tried to avoid letting into their cabs. I discovered that whereas most taxicab drivers in Kunming are those who were initially marginalized in China's reform era—laid-off factory workers and displaced farmers—the drivers now view as dangerous those people who are even more marginalized—the unemployed, the landless, the disaster-stricken, as well as certain minority groups whom they consider too poor or too troublesome. The cabdrivers label as dangerous those places that are liminal ones—what might be thought of as the limits, the boundaries and portals of the city—between urban and rural space, where marginalized people enter the city, or where new forms of consumption occur (see Collins, this volume, on the intersecting boundaries of gender, space, and cultural practices).

Kunming taxi drivers' *discourses of avoidance* share some of the same features as expressions of urban "racialized fears" in the Americas and Europe (Hesse 1997; Lowe 2001, 2003; McCallum 2005; Westwood and Williams 1997). (See Tang and Perry, this volume, on discussions that pertain to racialized fears.) However, the discourses of avoidance expressed by the drivers take on particular contours of fears of difference (of cultural, religious, place-based, physical, economic, gender difference) in the Chinese context. Whereas these discourses of avoidance often draw on earlier categories of undesirable Others in China, they take on new valences in the present environment of creative destruction, reconstruction and economic restructuring of the city.

Conversations with taxi drivers suggest that even those who have been socioeconomically marginalized in China's reform era are highly complicit in processes of Othering and boundary making. As anthropologist Helen Siu has recently asked: "How complicit is local society, victims included, in materializing this Othering process?" (Siu 2006: 331). Whereas Kunming's taxi drivers are physically mobile, they see this mobility as compromised by their low-status position. Rather than being sympathetic to those like themselves who have also been marginalized or displaced, many of the drivers actively seek to distance themselves and limit the mobility of similarly marginalized others, at least conceptually.

THE CITY OF KUNMING

Like other Chinese cities during the past decade of phenomenal economic growth, Kunming has been transformed from a low-rise city dominated by bicycles into a high-rise city of broad avenues jammed with cars and taxis. When I think of Kunming, what first comes to mind is the city where I lived and studied in the fall of 1993, a city of quiet tree-lined streets flowing with thousands of bicycles, a city of low, wood and stucco homes with tile roofs, a city with an old business district of French-inspired art deco buildings. But that city is gone. Gone are the old streets, the old homes, and most of the old buildings. In their place are broad, new car-friendly avenues, condominiums, skyscrapers and gleaming shopping malls. As anthropologist Zhang Li (2006) has observed, this massive destruction and reconstruction of Kunming has been accomplished through the collaboration (or one might say collusion) of officials and developers in a mad rush for modernity and for "catching up" with the coastal megacities of Beijing, Guangzhou, Shanghai and Shenzhen (for earlier transformations of Kunming see Gaubatz 1995, 1996; and Liu and Chen, this volume, on the rapid economic growth of Tianjin).

Despite being an inland provincial capital and a "smaller" Chinese city (of over six million), Kunming is an "intensely global" city (Smart and Smart 2003: 269–270). Part of the Greater Mekong Subregion (Chen 2005), Kunming is China's hub to Southeast Asia. From Kunming one can fly directly to Bangkok, Kuala Lumpur, Singapore, Mandalay, Yangon, Hanoi, Vientiane, Phnom Penh, as well as New Delhi and Seoul.

Not only has the physical space of the city changed radically in the reform era; in Kunming, the official, permanent, metropolitan area population has risen by 400,000 people in the last seven years, from 5.78 million in 2000 to 6.19 million in 2007 (see "Population Overview" 2008), in part because the city boundaries were extended outward. But the official population figures do not take into account the "floating population" of rural-to-urban migrants. Members of the "floating population" work at some of the city's most dangerous and dirty jobs: in demolition and construction for men and in domestic service, garbage sorting, begging and prostitution for women.

In the new physical, social and economic landscape of Kunming, how do people move through the city? How is their physical mobility related to their socioeconomic mobility? And how do they represent themselves and others in this new urban context? How do they imagine and narrate the city? One way for me to trace answers to these questions in contemporary Kunming has been to talk with taxi drivers.

TAXI DRIVERS IN KUNMING

When I returned to Kunming in January of 2008 I had not intended to conduct research with taxi drivers. My research initially focused on new middle-class car consumers. However, as I usually took taxis to get around to research sites (for example, to auto shows, car dealers and car modification shops), and because taxis are convenient and relatively inexpensive in U.S. terms, just eight yuan (a little over one U.S. dollar) for the first three kilometers; and because the taxi in which I rode was often stuck in traffic for over thirty minutes to an hour, I began to take an interest in taxi drivers, their work, and their views of the rapidly changing city.

From the end of January to mid-July 2008, I spoke with over sixty taxi drivers in Kunming. Even though I often caught a cab in one of two locations, I never had the same cabdriver twice (as in the United States, a driver's name and photograph are posted above the dashboard for passengers to clearly see). This may be due to the fact that in Kunming there are no zone regulations, so a cabdriver can pick up a fare anywhere in the city if he or she chooses, and drive to anywhere, again if he or she chooses. My conversations were limited to those who drove officially licensed cabs, as distinguished in Kunming by their turquoise blue color. Usually these cabs were a VW sedan, either a Jetta or a Passat.

Taxicabs in Mandarin are formally called *chuzu qiche*, "cars for hire." Informally, cabs are called *di*, short for *dishi*, from the Cantonese transliteration of taxi, *diksee* (Barmé 2002: 186). One calls a cabdriver "master" (*shifu*), a term of respect often used for those in skilled service positions. Before the start of China's reform era in 1978, only Chinese political elites and foreigners rode in cars and cabs, and to become a driver, one had to attend a driving school full time for one year. Now in Kunming, someone can become a cabdriver if he or she has three years' driving experience, no record of liability in traffic accidents; a city residence permit; is under age sixty for men and under age fifty-five for women, has passed a taxi training course; is deemed capable of "civil" conduct; and as of October 11, 2008, had a junior high school education or above ("Taxi Driver" 2008).

Unlike in New York, where 99 percent of the drivers are now men (Design Trust, 2007: 50), in Kunming women constitute a sizable percentage of taxi drivers.[2] Although the Yunnan Traffic Bureau does not as yet keep any detailed statistics on taxi drivers, one in five of the drivers with whom I spoke were women. The high percentage of women taxi drivers seems unusual for other places in China for it is often remarked upon by visitors to the city. Whereas I have never had a woman taxi driver in another Chinese city, Berlin-based translator Fang Yu's fascinating documentary film *Women at the Wheel*, about the lives of three women taxi drivers in the city of Xi'an, and anthropologist Emily Chao's article (2003) discussing rumors about women taxi drivers in the northwestern Yunnan town of Lijiang, suggest that there are at least some women drivers elsewhere in China. However, surprisingly, as I will discuss below, taxi driver perceptions of dangerous people and places cut across the gender of the drivers, with one exception, and that is avoidance of single women passengers. Whereas some taxi drivers in Kunming are Muslim (as distinguished by

Arabic inscriptions across their windshields), the drivers of the cabs I happened to catch identified themselves as Han Chinese.

As in New York, in Kunming, "[s]ignificant differences in class exist among cabdrivers" (Hodges 2007: 4). In Kunming, drivers either own their cabs, or rent them, either from a company or from an individual owner. Those who own their cabs can earn five times more than those who rent. If a driver owns his or her cab, his or her income is much higher, 8,000 yuan to 10,000 yuan per month (approximately $1,000 to $1,250 dollars) than if he or she rents, whereupon a driver's income is between 1,800 to 2,000 yuan per month ($225 to $250 dollars per month). Still, even the income of those who rent cabs is double that of the average per capital urban income, which is less than 1,000 yuan per month; and eight and a half times higher than the average per capita rural income, which is only 220 yuan per month ($27.50) (*Yunnan sheng tongji ju* 2008). Although drivers earn above the urban per capita average income, they referred to themselves as part of the *laobaixing*; literally "the old hundred names," meaning the "regular" people, as distinguished from political, economic and cultural elites (Notar 2009).

Kunming taxi drivers drive six days per week with one mandated day off per week. In Kunming, as in New York City, two drivers share a cab, alternating twelve-hour shifts. As one driver told me, "We sleep, but the taxi never sleeps." Drivers in Kunming often complained to me of the exhaustion of sitting and driving in heavy traffic for twelve hours: "All I want to do when I get home is sleep" was a common refrain. Drivers who lease their cabs are under intense pressure to garner enough fares to meet their lease. As Biju Mathew, in his book *Taxi!*, about New York cabdrivers, describes, the lease sits "around [the driver's] neck like a noose" (Mathew 2005: 138). Drivers who own their cars are under less pressure, but have often taken loans from friends or family to buy their cars and must repay those loans with interest. When working, drivers rarely get out of their cabs except to go to the bathroom. I noticed that drivers would often just eat a bowl of instant noodles for lunch while sitting in their cabs. For the drivers, their cabs become a kind of "iron cage" to which they are confined.

Of the Kunming drivers with whom I spoke, most had had economic difficulties in the reform era. Many were former factory workers who had either been laid off (*xiagang le*), had quit, or were still formally on the books as working at a particular factory, but in reality received no salary and did no work for the factory. These former factory workers described their driving a cab as a kind of shameful "coming out" from the factory system to "hit the road" (*mei banfa jiu chulai shanglu*). Some were former farmers who had either lost their land or had someone else back home who could work it. Some had been former truck drivers who had either been laid off or who had quit to stop their long-distance travel.

Despite making good money, the shame associated with driving a cab seemed to come from a sense of downward social mobility. Those who had been former factory workers or truck drivers considered driving a cab of lower status, for they had entered the realm of service work (associated with the more feminized, low-paid jobs).

SCHOLARLY WORK ON TAXIS

Surprisingly, given many taxicab drivers' propensity for conversation, social science literature on taxis has been sparse (see Davis 1959; Gilbert and Samuels 1982; Lawuyi 1988; Stoller 1989; Verrips and Meyer 2000; also Leonard 2006). Recently, Biju Mathew, in *Taxi!* (2008 [2005]), has documented the struggles and strikes of the New York Taxi Workers' Alliance, which he helped organize. Diego Gambetta and Heather Hamill, in *Streetwise: How Taxi Drivers Establish Their Customers' Trustworthiness* (2005), come closest to what I seek to do here in examining taxi-driver perceptions of others, although they are more interested in the psychological aspects of risk perception than the sociocultural imaginations of who is risky or to be avoided.

In China in particular, scholarly interest has focused on how taxi drivers have been viewed by others. In an essay on the history of the car in China, Geremie Barmé has written briefly about taxis in reform-era China, noting that the taxi driver "became one of the most liminal figures in the urban landscape," because taxis "ferried prostitutes . . . and provided the covert environment for illicit contacts of all descriptions" (Barmé 2002: 185). Emily Chao describes rumors about women taxi drivers in the tourist town of Lijiang, located in northern Yunnan province. In local gossip about the untimely deaths or reputedly "stinky" bodies of the women taxi drivers (indicating that they may be potential fox spirits), Chao finds allegories of "how capitalist privatization [has been] straining the social fabric of Lijiang" (Chao 2003: 72).

Inspired by Barmé's and Chao's writings on views of taxi drivers in Chinese cities as liminal and suspect figures, I wondered how taxi drivers themselves view others. I have always enjoyed talking with taxi drivers in China. Because taxi drivers work alone in the enclosed bubble of their vehicle—a strange combination of a private, bounded bubble moving through highly public space—they will often speak frankly about a range of topics.

TAXICAB DRIVERS' VIEWS OF DANGEROUS PLACES AND PEOPLE

In the spring of 2008, as I spoke with taxi drivers, I began to ask them two questions: First, is there anywhere in the city they avoid going? And second, is there anyone they avoid letting ride in their cabs? I sought to understand their geographic imaginary of the city as they moved through it.

Dangerous Places

"Is there anywhere you do not like to go or do not dare to drive?" In response to this question, many of the drivers (across ages, genders, and incomes) told me that they did not like to drive outside of the city, specifically beyond the second ring road, day or night. They considered the realm beyond that ring road to be dangerous. For example, some of the male drivers said they feared that even if they took a single woman passenger outside the second ring road, she might have arranged for a group of men to be waiting in ambush to rob him. In a recent article on the city of Guangzhou, anthropologist Helen Siu has observed that urbanites there view the village "enclaves" on the outskirts of the city, which have been "swallowed up" by the city but are not considered within the city proper, "as sites of crime and disease, cancers that threaten a modern cityscape" (Siu 2006: 330). Kunming taxi drivers held a similar view of these liminal spaces that are neither city nor country. In contrast to many Anglo-American middle-class perceptions of the city as a place of danger and the suburbs as a place of safety, Kunming cabdrivers held the reverse view, more akin to a central Parisian view of the suburban *banlieues* as troubled outer spaces.

The second type of place some drivers tried to avoid was the stations. Like the boundaries between the city and its outskirts, the bus and train stations are also liminal spaces, portals between the city and elsewhere. For example, one driver told me that she did not like to go to the stations because this is where migrants would enter the city and she was afraid that they might try to steal her jewelry. This too is similar for what Siu observed in Guangzhou (Siu 2006). Yet, no taxi drivers mentioned trying to avoid the airport, a similarly liminal space, but one that is dominated by elite travelers—businesspeople, officials, and national and transnational tourists.

The third type of place that some drivers told me they tried to avoid was one of the new centers of consumption in the city: Walmart. This I found surprising because it did not fit the pattern of the other two places as ones associated with migrants. Yet one driver explained his avoidance of Walmart in terms of its association with outsiders: "I avoid going to Wal-Mart and you should too," he warned me. "There are too many people from Xinjiang who hang out around there."

I must admit that I went to the main Walmart in Kunming more times than I care to mention, especially when my family and I first arrived and I needed to furnish the apartment we rented. To my dismay, the old Kunming department store no longer existed, and the new department store did not carry household items. When I asked where I could find a rice cooker, wok, teakettle, bedding, etc., the consensus among the university faculty and staff was: Walmart.

Yet I never noticed anyone identifiably from Xinjiang in or around Walmart. I assumed that the cabdrivers were using "people from Xinjiang" (*Xinjiang ren*) not to refer to Han Chinese migrants to Xinjiang but as a shorthand for ethnic Uyghurs from Xinjiang who might be distinguished by curlier hair, hazel eyes, and/or their attire, caps for men (*kufi*, like yarmulkes, or squat, embroidered *doppa*) and headscarves (*hajab*) for women. The customers I observed at Walmart were mainly middle-class Chinese (unlike in the United States, Walmart in China is geared more toward middle- and upper middle-class consumers) and a few foreigners—students and teachers—like myself. It is possible that some of these European and American foreigners were mistaken as *Xinjiang ren*—with my curly brown hair and hazel eyes I was sometimes asked if I were from Xinjiang. However, I suspect that Walmart more generally had been marked as a place of "outside" and "outsider-ness" where new consumer goods and middle-class consumption practices had entered the city. It was a liminal space of consumption that had been racially marked as a site where potentially dangerous outsiders (or foreigners) might lurk.

Whereas elites interviewed by anthropologists in other cities, for example, São Paulo and New York, fear the city as a place of danger (Caldeira 2001; Lowe 2001, 2003), the taxi drivers with whom I spoke expressed avoidance of the boundaries between city and country, the new suburbs of the city, the "'restless urban frontier'" (McGee et al. 2007: 72–73). They also spoke of trying to avoid other liminal places—the bus and train stations (where migrants could enter the city) and Walmart (marked as a site of "outsider-ness").

Dangerous Peoples

In addition to wanting to know where the taxi drivers tried to avoid going, I was also curious about whom they might try to avoid having in their cabs. Using what Diego Gambetta and Heather Hamill, in their book *Streetwise: How Taxi Drivers Establish Their Customers' Trustworthiness* (2005), call "screening"—looking at someone's appearance, and "probing," talking with someone to try to ascertain what kind of person this was before allowing him or her into the cab—taxi drivers in Kunming consistently mentioned four types of passengers to be avoided. These can be grouped by place of origin, ethnicity, gender and occupation.

The first type of "undesirables" Kunming taxi drivers mentioned were categorized by place of origin. Of these, drivers commonly noted trying to avoid people from Sichuan province, who drivers tried to distinguished from Yunnanese (*Yunnan ren*—Han Chinese from Yunnan) by their accents, with the question "Where do you want to go?" Drivers referred to *Yunnan ren* as generally "honest" folk (*laoshi*), whereas they called people from Sichuan "rats" (*Sichuan haozi*)—and described them as "sly" and "cunning." Although people in Kunming overall expressed sympathy for those in Sichuan who had suffered in the massive earthquake on May 12, 2008, they were also worried that displaced Sichuanese were going to invade Kunming like a horde of rats.

Whereas drivers preferred passengers from Yunnan, within the category of "people from Yunnan," twelve of the drivers mentioned that they would try to avoid people from northeast Yunnan (*Dian dongbei*), the lobe of Yunnan that sticks out into Sichuan and Guizhou provinces. Northeast Yunnan is a poor place associated with the Yi, Hui (Muslim) and Bai minority nationalities (Zhang 2003: 38–41). Drivers particularly mentioned trying to avoid people from two places: Zhaotong and Zhenxiong. These two places, like Sichuan province, have been disaster-stricken areas. Over the past decade, Zhaotong and Zhenxiong have been the sites of repeated mudslides and landslides that have destroyed hundreds of homes and displaced thousands of people. When I asked one driver how he could tell if someone was from this area, he laughed and said, "Well, they do not wear signs around their necks saying 'I am from Zhaotong'! I have to listen for their local accent." Other drivers told me that the people from this area had "black teeth" (*hei yachi*). When I asked why, the drivers speculated that the food and water there must be bad. Whereas some taxi drivers singled out people from Zhaotong and Zhenxiong as being marked by physical difference, this difference is not unique to those areas. Many rural men in Yunnan have brown teeth due to smoking low-grade unfiltered cigarettes and drinking strong, earthy tea. (In fact, when I lived in villages in northwest Yunnan from 1994 to 1995 and drank strong tea, I also had *hei yachi*!). In other words, whereas a certain physical marker may be used to stigmatize groups, the marker does not necessarily correspond solely to a targeted group.

As geographer Tim Cresswell has argued: "The definition of insider or outsider is more than a locational marker. . . . An outsider is not just someone literally from another location but someone who is existentially removed from the milieu of 'our' place" (Cresswell 1996: 154). In avoiding people from Sichuan and northeast Yunnan, drivers seemed to want to avoid those in desperation. Whether out of fear, guilt or both, many of the drivers did not want to ferry those they imagined as displaced or disaster-stricken

The second type of passenger drivers told me they tried to avoid was ethnically marked "Others": either Muslims (*Hui zu*), sometimes specifically referred to as people from Xinjiang, and the Hmong/Miao (*Miao zu*) from Guizhou province. Drivers described the Muslims as "dangerous" and "troublemakers." The Miao, on the other hand, were "poor" and "dirty."

I was particularly interested in the taxi drivers' focus on the Muslims and the Miao, for it highlighted who was *not* mentioned. In mid-March of 2008 when I was talking with drivers, Tibetan demonstrations in China were met with police crackdowns and then more demonstrations. The northern part of Yunnan province, a scenic Tibetan area, was under lockdown, and closed off to both national and transnational tourists. After mid-March, more security personnel appeared on the streets of Kunming, and a large international anthropology conference was canceled, for what seemed to be security concerns. Yet none of the taxi drivers ever mentioned Tibetans as potentially dangerous; neither did they mention any of the twenty-two other minority nationalities (*shaoshu minzu*) of Yunnan.

In addition, in the spring of 2008, I noticed many more Burmese migrants in Kunming working as petty street traders. They were mostly from the city of Mandalay, and of Chinese-Burmese descent. Smuggled goods (including heroin) enter China from Burma (Myanmar) through Yunnan. But none of the drivers mentioned Burmese as potentially dangerous passengers to be avoided.

Instead, taxi drivers singled out Muslims and Miao. Interestingly, both of these groups are older, ethnically marked categories of "dangerous peoples" who had rebelled in the nineteenth century. After the Yunnan Panthay Rebellion (1856–1873) was brutally suppressed by Qing imperial forces, the Yunnan Muslims were scapegoated and their property confiscated, even though different groups (Han, Bai, Yi) had fought on both sides of the battle (Atwill 2005; Notar 2001). Also during the mid-nineteenth century, a rebellion in neighboring Guizhou province was referred to as the *Miao luan*, literally the "Miao chaos," even though many groups participated (Jenks 1994: 3–4). Long before this, the Miao had been depicted condescendingly as "primitive barbarians" (Deal and Hostetler, trans. 2006; Diamond 1988; Jenks 1994; Schein 2000; Tapp 2001).

What is interesting is that the category of these two "problem" groups from the mid-nineteenth century have been reinvigorated and singled out as creating problems in the present. In her history of Shanghai residents' bias against people from Subei (*Subei ren*), Emily Honig points out that the bias against *Subei ren*, or anyone suspected of being from Subei, has persisted for over a century. Honig traces this bias to Shanghainese unwillingness to accept Subei peasants who were fleeing natural disaster in the late nineteenth and early twentieth centuries—an attitude that resonates with current Kunming taxi drivers' professed avoidance of Sichuanese and people from northeast Yunnan who have been fleeing disaster-stricken areas. However, whereas Honig documents that Shanghainese prejudice against *Subei ren* has persisted, she urges us to keep in mind that the meanings and valences of this category have changed over time to incorporate new circumstances (Honig 1992).

Similarly, we might consider that whereas "Muslims" and "Miao" have long been categorized as dangerous Others in China's southwest, the particular valences associated with these categories can change over time (see Gillette 2000; Gladney 1991, 1994; Harrell 1995; Schein 2000; Tapp 2001). In the reform era, Yunnan Muslims, with a long history as traders and entrepreneurs, are suspected of dominating the current heroin trade (although posters listing the names of people executed for smuggling suggests that the trade is multiethnic). And in a post-9/11 world, Chinese Muslims, along with Xinjiang Uyghurs (who Kunming cab drivers referred to as *Xinjiang ren*), are suspected of religious extremism and terrorism. Miao, on the other hand, are no longer described as rebels or troublemakers, as they were in the nineteenth century, but simply impoverished, engaging in "dirty" activities like garbage collection and begging (although it is unlikely that a garbage collector or beggar would try to hail a cab).

The third group of passengers several male taxi drivers told me that they were wary of, but could not avoid altogether, were single, beautiful women. Some of these women, the male drivers told me, would sit in the front passenger seat instead of in the rear. At some point during the ride, these women would offer the driver a piece of candy or a cigarette. However, these were not normal candies or cigarettes—they would be laced with a sedative (*anmiyao*) or poison (*du*). After a driver consumed them or smoked them, he would fall into a deep sleep. When he awoke several hours later, he would find that all of his money had been stolen.

Of all the stories about potentially dangerous passengers that I heard from drivers, these were the most like urban legends. They seemed to serve as morality tales about temptation—of having a single woman sitting too close and of letting down one's guard. In the stories that I heard, drivers did not mark the women as being ethnically "other" but the tales were reminiscent of other stories that circulate (and have circulated for centuries) in Yunnan about Miao women who will poison Han men with poison-pot poison (*gu*) and seduce them (see Diamond 1988). These tales also served as moral commentaries on women who were considered too bold or independent, i.e., that such women can only be dangerous (see Chao 2003).

The fourth category of passengers taxi drivers said that they tried to avoid was occupational. When one driver first told me that he tried to avoid picking up soldiers in his cab, I could not help but laugh. "Aren't soldiers one of the three pillars of the People (*renmin*)—the workers, the peasants and the soldiers (*gong, nong, bing*)?" I joked. "Well, maybe they used to be," he laughed too, "but now . . . soldiers usually come from poor farm families. They are stuck in camps outside of Kunming, and only get to come into the city once in awhile. When they try to catch a cab to take them back to their camp, you have to drive outside the city. And then, these guys have no money, maybe just a couple of cigarettes. What are you going to do if they don't pay? Start a fight with them?" Rank-and-file soldiers in China are like migrants in the sense that they have left the countryside in search of opportunity. The soldiers were also located in liminal spaces, in camps at the edge of the city.

Other scholars have examined the prejudices of urbanites toward new rural-to-urban migrants in China (e.g., Yan 2008; Zhang 2001a, 2002). The Kunming taxi drivers (although themselves not all long-time Kunming residents) hold similar views. Yet, what I found interesting was that not all migrants are similarly disparaged. Certain migrants in particular are perceived to be more dangerous than others due to their place of origin, their nationality, their religion, their gender and social status, and their occupation.

The drivers' discourses of avoidance of certain places and peoples point to a greater "panic of mobility" (O'Dell 2001) in the city. Whereas there has long existed a two-tiered hierarchy in China between those who have urban citizenship and those who have rural citizenship, the reform era has introduced a more complex hierarchy within the urban (Zhang 2001a, 2002). During the Mao years (1949–1976), while political elites lacked disproportionate purchasing power and while available consumer goods were limited (Davis 2000), elites *were* allocated preferential access to goods and services based on rank: they had better housing, better food, and better transportation (chauffeured limousines or cars). With the reform era (1978 to the present), political elites as well as a new generation of economic elites (bankers, industrialists, entrepreneurs), and a growing white-collar middle class (*zhongchan jieceng*; *gongxin jieceng*) can purchase their own homes in new gated communities and their own cars, and indulge in new types of leisure experiences like trips abroad and memberships in golf clubs (Notar 2006, 2009; Zhang 2009; Zhang 2008, 2010). The taxi drivers, on the other hand, are those who have been marginalized in the reform era—laid-off workers or displaced farmers. Due to their experiences, one might expect them to be more sympathetic to those who have been similarly marginalized. However, I found on the contrary that Kunming taxi drivers were actively involved in maintaining urban boundaries that at least conceptually limit the mobility of others.[3]

CONCLUDING THOUGHTS

In *Landscapes of Power*, Sharon Zukin suggests that anthropologist Victor Turner's concept of liminality, first discussed in the context of being 'betwixt and between' social roles during ritual, is the most useful for understanding postmodern demolished and rebuilt urban landscapes. She writes: "The only possible perspective on this process of creative destruction is one of liminality" (Zukin 1991: 28; see Turner 1967).

In an overarching analysis, applying the concept of liminality to the current Chinese urban context works well. First, temporally, people are 'betwixt and between' three decades of revolutionary socialist modernism under Mao and a new as-of-yet-in-process system, which is still being defined. Variously called "socialism with Chinese characteristics," "late socialism," "post-socialism," or "Chinese neoliberalism," it is an increasingly privatized economy, yet one firmly orchestrated by the Chinese Communist Party (Zhang and Ong 2008). Due to this contradiction there exists a temporal, practical and ideological state of liminality.

Spatially, it is difficult to imagine any more massive "creative destruction" and reconstruction than that which has happened in Chinese cities over the past decade. It would be as if *most* of New York, Los Angeles, Chicago, Boston, San Francisco, Seattle, Miami, Phoenix, Washington, DC, and all other U.S. cities had been demolished and then rebuilt within ten years. Urban residents of Chinese cities have had to adjust quickly to entirely new cityscapes. Older fixed points of reference no longer exist: that neighborhood, that street, that building are gone. Chinese urbanites exist in a state of spatial and material liminality.

Perhaps it is precisely because of this temporal, ideological, spatial and material liminality that Chinese urbanites such as taxi drivers in the city of Kunming are attempting to impose conceptual boundaries on places and peoples (see Collins, this volume, on how similar dynamics played out in a very different spatial context). Whereas taxi drivers have been viewed by others as dangerously liminal figures—endlessly traversing the city, seemingly everywhere and nowhere, providing a space for illicit encounters (Barmé 2002; Chao 2003)—they themselves, although marginalized figures—mostly laid-off workers and displaced farmers—actively circumscribe the mobility of even more marginalized persons: the disaster-stricken, displaced and impoverished; certain minority nationalities; overly independent single women; and poor soldiers.

Whereas the concept of liminality works well for thinking temporally of China as a nation-state betwixt and between socialism and something else, and Chinese cities as spatially betwixt and between urban forms, it works less well if we think of different groups within cities. Turner used the concept of liminality to refer to a group of initiates in ritual who were between one social status and the next. Whereas taxi drivers in Kunming have moved from one form of employment to another—mostly from factory work to service work—they are not without status. It is this new, lower social status that perhaps motivates many of them to draw boundaries between themselves and those who are even more marginalized.

Anthropologist Helen Siu has suggested that in the contemporary Chinese context of transformation and rural-to-urban migration, both "material and conceptual boundaries are in flux" (Siu 2006: 333). Whereas material and ideological boundaries are radically in flux, it seems, surprisingly, that other conceptual boundaries are not in flux in the same way. Instead, when asked where it is dangerous and who is dangerous, taxi drivers responded by drawing on preexisting categories: "outside the city," "the peasant," "the foreign" (Walmart), "the single woman," "the Muslim," and the "Miao." These familiar categories seem to represent an effort to try to assert an older conceptual order on the radically new physical and social landscapes of the city, to harness old categories for new conditions (see Moser, this volume, on the reassertion of traditional architectural forms in the new Malaysian and Indonesian cities to stand for a new national identity in a more global and fluid environment).

Moreover, the categories that the taxi drivers used are local, or at least regional, not national categories. At a time of Tibetan unrest, none of the taxi drivers mentioned Tibetans as dangerous, only the Muslims and the Miao, regional historic "troublemakers." Conversely, after the massive earthquake of May 2008, when there existed national goodwill for the people of Sichuan, Kunming taxi drivers voiced derogatory views of

Sichuan migrants. And whereas in the national media a discourse suggesting the potential for the "quality" (*suzhi*) of even assumedly "backward" peoples such as rural migrants to be "improved" (see Kipnis 2006: 296–297), the taxi drivers spoke of the Sichuanese, Muslims, Miao and people from northeast Yunnan as innately undesirable. That migrants came from a particular place or that they were of a particular ethnicity made them *inherently* problematic. Whereas the discourses of avoidance expressed by taxi drivers in Kunming share "hypothetical equivalence" (Liu 1995) with the racialized fears expressed by elites in New York (see Lowe 2001, 2003), they are not coterminous with them. We must attempt to understand the specific local contours of bias.

China, we might say, is now becoming a capitalist place, but the conceptual imaginaries the taxicab drivers employ date to a precapitalist era. The stereotypical images evoked by the cabdrivers—of the countryside and farmers in China as both backward and dangerous and the city as civilized and safe—seem contrary to English capitalist-conceived notions of the pastoral and the industrial. This suggests that we need to examine not only what Raymond Williams referred to as the "persistence and historicity of concepts" (Williams 1973: 289) closely within particular places, at particular moments, but also not assume that all urban imaginaries perpetuated under market transformations will be globally or even nationally similar.

ACKNOWLEDGMENTS

Revised paper presented at the conference on "Rethinking Cities and Communities: Urban Transition before and During the Era of Globalization," sponsored by The Center for Urban and Global Studies, Trinity College, Hartford, Connecticut, November 14–15, 2008; also presented at the American Anthropological Association Annual Meetings in San Francisco, November 22, 2008, and for the panel Mobility Mentalities: Moving through Urban Spaces in Asia at the Association for Asian Studies Annual Meetings in Chicago, March 27, 2009. Thank you to Xiangming Chen for inviting me to participate in the Trinity conference and to Joshua Hotaka Roth for organizing the other two panels. I am especially grateful to Emily Chao and to Helen Fung Har Siu, discussants for these panels, who provided excellent questions and suggestions for revision. I have also benefited from stimulating discussion with copanelists Joshua Hotaka Roth, Thomas Williamson and Jun Zhang. Research was funded through a National Endowment for Humanities Summer Stipend and a Trinity College Faculty Research Grant.

NOTES

1. Traffic congestion has increased in Kunming, as in other Chinese cities, as personal car consumption has increased. Research for this chapter forms

part of a book project: *Autobiographies: Narratives of Cars and Mobility in Contemporary Urban China.*

2. There were numerous women cabdrivers in New York and other U.S. cities during WWII; see Hodges 2007.
3. The focus here is on drivers' *conceptions* of dangerous people and places. Future research might examine whether certain marked passengers do in fact have difficulties hailing a cab.

REFERENCES

Atwill, David G. (2005). *The Chinese Sultanate: Islam, Ethnicity, and the Panthay Rebellion in Southwest China, 1856–1873.* Stanford, CA: Stanford University Press.

Barmé, Geremie R. (2002). "Engines of Revolution: Car Cultures in China." Pp. 177–190 in *Autopia: Cars and Culture*, edited by Peter Wollen and Joe Kerr. London: Reaktion Books.

Caldeira, Teresa (2001). *City of Walls: Crime, Segregation, and Citizenship in São Paulo.* Berkeley: University of California Press.

Chao, Emily (2003). "Dangerous Work: Women in Traffic." *Modern China* 29: 71–107.

Chen, Xiangming (2005). *As Borders Bend: Transnational Spaces on the Pacific Rim.* Lanham, MD: Rowman & Littlefield.

Cresswell, Tim (1996). *In Place/Out of Place: Geography, Ideology, and Transgression.* Minneapolis and London: University of Minnesota Press.

Davis, Deborah S. (2000). "Introduction: A Revolution in Consumption." Pp. 1–20 in *The Consumer Revolution in Urban China*, edited by Deborah S. Davis. Berkeley: University of California Press.

Davis, Fred (1959). "The Cabdriver and His Fare: Facets of a Fleeting Relationship." *The American Journal of Sociology* 65 (2): 158–165.

Davis, Mike (1990). *City of Quartz: Excavating the Future in Los Angeles.* London: Verso.

Deal, David M., and Laura Hostetler, trans. (2006). *The Art of Ethnography: A Chinese "Miao Album."* Seattle: University of Washington Press.

Design Trust for Public Space and the New York City Taxi and Limousine Association (2007). *Taxi 07: Road Forward.* City of New York.

Diamond, Norma (1988). "The Miao and Poison: Interactions on China's Southwest Frontier." *Ethnology* 27 (1): 1–25.

Donald, James (1997). "This, Here, Now: Imagining the Modern City." Pp. 181–201 in *Imagining Cities: Scripts, Signs, Memory*, edited by Sallie Westwood and John Williams. London and New York: Routledge.

Fang Yu (2006). *Women at the Wheel* (Die Taxischwestern von Xian; in Mandarin, Di Jie). Film und Video Untertitelung Gerhard Lehmann AG.

Gambetta, Diego, and Heather Hamill (2005). *Streetwise: How Taxi Drivers Establish Their Customers' Trustworthiness.* New York: Russell Sage Foundation.

Gaubatz, Piper Rae (1995). "Urban Transformation in Post-Mao China: The Impacts of the Reform Era on China's Urban Form." Pp. 28–60 in *Urban Spaces in Contemporary China*, edited by Deborah S. Davis, Richard Kraus, Barry Naughton, and Elizabeth Perry. Washington, DC, and New York: Woodrow Wilson Center Press and Cambridge University Press.

Gaubatz, Piper Rae (1996). *Beyond the Great Wall: Urban Form and Transformation on the Chinese Frontiers.* Stanford, CA: Stanford University Press.

Gilbert, Gorman, and Robert E. Samuels (1982). *The Taxicab: An Urban Transportation Survivor.* Chapel Hill and London: The University of North Carolina Press.

Gillette, Maris Boyd (2000). *Between Mecca and Beijing: Modernization and Consumption among Urban Chinese Muslims.* Stanford, CA: Stanford University Press.

Gladney, Dru C. (1991). *Muslim Chinese: Ethnic Nationalism in the People's Republic.* Cambridge, MA: Council on East Asian Studies, Harvard University.

Gladney, Dru (1994). "Representing Nationality in China: Refiguring Majority/ Minority Identities." *Journal of Asian Studies* 53 (1): 92–123.

Harrell, Stevan (1995). "Introduction: Civilizing Projects and the Reaction to Them." Pp. 3–36 in *Cultural Encounters on China's Ethnic Frontiers*, edited by Stevan Harrell. Seattle: University of Washington Press.

Harvey, David (1985). *Consciousness and the Urban Experience: Studies in the History and Theory of Capitalist Urbanization.* Baltimore: Johns Hopkins University Press.

Hesse, Barnor (1997). "White Governmentality: Urbanism, Nationalism, Racism." Pp. 86–103 in *Imagining Cities: Scripts, Signs, Memory*, edited by Sallie Westwood and John Williams. London and New York: Routledge.

Hodges, Graham Russell Gao (2007). *Taxi!: A Social History of the New York City Cabdriver.* Baltimore: Johns Hopkins University Press.

Honig, Emily (1992). *Creating Chinese Ethnicity: Subei People in Shanghai, 1850–1980.* New Haven. CT: Yale University Press.

Jenks, Robert D. (1994). *Insurgency and Social Disorder in Guizhou: The 'Miao' Rebellion, 1854–1873.* Honolulu: University of Hawai'i Press.

Kipnis, Andrew (2006). "*Suzhi*: A Keyword Approach." *The China Quarterly* 186: 295–313.

Lawuyi, Olatunde Bayo (1988). "The World of the Yoruba Taxi Driver. An Interpretive Approach to Vehicle Slogans." *Africa* 58 (1): 1–13.

Leonard, Robert (2006). *Yellow Cab.* Albuquerque: University of New Mexico Press.

Liu, Lydia H. (1995). *Translingual Practice: Literature, National Culture, and Translated Modernity—China, 1900–1937.* Stanford, CA: Stanford University Press.

Lowe, Setha M. (2001). "The Edge and the Center: Gated Communities and the Discourse of Urban Fear." *American Anthropologist* 103 (1): 45–58.

Lowe, Setha (2003). *Behind the Gates: The New American Dream? Searching for Security in America.* New York: Routledge

Mathew, Biju (2005). *Taxi! Cabs and Capitalism in New York City.* Ithaca, NY: Cornell University Press.

McCallum, Cecilia (2005). "Racialized Bodies, Naturalized Classes: Moving through the City of Salvador da Bahia." *American Ethnologist* 32 (1): 100–117.

McGee, Terence G., George C. S. Lin, Andrew M. Marton, Mark Y. L. Wang, and Jiaping Wu (2007). *China's Urban Space: Development under Market Socialism.* London and New York: Routledge.

Notar, Beth E. (2001). "Du Wenxiu and the Politics of the Muslim Past." *Twentieth-Century China* 26 (2): 63–94.

Notar, Beth E. (2006). *Displacing Desire: Travel and Popular Culture in China.* Honolulu: University of Hawai'i Press.

Notar, Beth E. (2009). "Mixed Mobilities: Taxi Drivers and Day Trippers in Southwest China." Paper presented to the Yale University East Asia Anthropology Lunch Lecture Series. New Haven, April 3.

O'Dell, Tom (2001). "Raggare and the Panic of Mobility: Modernity and Everyday Life in Sweden." Pp. 105–132 in *Car Cultures*, edited by Daniel Miller. New York: Berg.

"Population Overview"(2008). Kunming, China, Basic Facts of Kunming. August 31. http://en.km.gov.cn/showArticle.aspx?cid=23&aid=288; accessed October 3, 2008 (Kunming government Web site).

Pun, Ngai (2005). Made in China: Women Factory Workers in a Global Marketplace. Durham, NC: Duke University Press.

Schein, Louisa (2000). *Minority Rules: The Miao and the Feminine in China's Cultural Politics*. Durham, NC: Duke University Press.

Siu, Helen. F. (2006). "Grounding Displacement: Uncivil Urban Spaces in Post-Reform South China." *American Ethnologist* 34 (2): 329–350.

Smart, Alan, and Josephine Smart (2003). "Urbanization and the Global Perspective." *Annual Review of Anthropology* 32: 263–285.

Solinger, Dorothy J. (1999a). *Contesting Citizenship in Urban China*. Berkeley: University of California Press.

Solinger, Dorothy (1999b). "China's Floating Population." Pp. 220–240 in *The Paradox of China's Post-Mao Reforms*, edited by Merle Goldman and Roderick MacFarquhar. Cambridge, MA: Harvard University Press.

Stoller, Paul (1989). *The Taste of Ethnographic Things. The Senses in Anthropology*. Philadelphia: University of Pennsylvania Press.

Tapp, Nicholas (2001). *The Hmong of China: Context, Agency, and the Imaginary*. Leiden, Boston: Brill.

" 'Taxi driver' in Kunming Requires Education above Junior Secondary School" (2008). Oct.11. Kunming, China, *Kunming News*. http://en.km.gov.cn/showArticle.aspx?cid=57&aid=402; accessed 31 October 2008.

Turner, Victor (1967). *The Forest of Symbols: Aspects of Ndembu Ritual*. Ithaca, NY: Cornell University Press.

Verrips, Jojada, and Birgit Meyer (2000). "Kwaku's Car: The Struggles and Stories of a Ghanaian Long-Distance Taxi-Driver." Pp. 153–184 in *Car Cultures*, edited by Daniel Miller. Oxford: Berg.

Westwood, Sallie, and John Williams (1997). "Imagining Cities." Pp. 1–16 in *Imagining Cities: Scripts, Signs, Memory*, edited by Sallie Westwood and John Williams. London and New York: Routledge.

Williams, Raymond (1973). *The Country and the City*. New York: Oxford University Press.

Yan, Hairong (2008). *New Masters, New Servants: Migration, Development, and Women Workers in China*. Durham, NC: Duke University Press.

Yunnan sheng tongji ju (2008). "Yunnan sheng 2007 nian guomin jingji he shehui fazhan tongji gongbao" (Statistical report on Yunnan province 2007 people's economic and social development). Yunnan sheng dianzi zhengwu menhu wangzhan (Yunnan e-government) July 29. http://www.yn.gov.cn; accessed 25 October 2008.

Zhang, Jun (2009) Driving Toward Modernity: an Ethnography of Automobiles in Contemporary China. Yale University dissertation.

Zhang, Li (2001a). *Strangers in the City: Reconfigurations of Space, Power, and Social Networks within China's Floating Population*. Stanford, CA: Stanford University Press.

Zhang, Li (2001b). "Contesting Crime, Order, and Migrant Spaces in Beijing." Pp. 201–222 in *China Urban: Ethnographies of Contemporary Culture*, edited by Nancy Chen et al. Durham, NC: Duke University Press.

Zhang, Li (2002). "Spatiality and Urban Citizenship in Late Socialist China." *Public Culture* 14 (2): 311–334.

Zhang, Li (2006). "Contesting Spatial Modernity in Late-Socialist China." *Current Anthropology* 47 (3): 461–476.
Zhang, Li (2008). "Privatizing Homes, Distinct Lifestyles: Performing a New Middle Class." Pp. 23–40 in *Privatizing China: Socialism from Afar*, edited by Li Zhang and Aihwa Ong. Ithaca, NY: Cornell University Press.
Zhang, Li (2010). *In Search of Paradise: Middle-Class Living in a Chinese Metropolis*. Ithaca, NY: Cornell University Press.
Zhang, Li, and Aihwa Ong (2008). "Introduction: Privatizing China: Powers of the Self, Socialism from Afar. " Pp. 1–19 in *Privatizing China: Socialism from Afar*, edited by Li Zhang and Aihwa Ong. Ithaca, NY: Cornell University Press.
Zhang, Ning (2003). *Zhaotong*. Yunnan xiangtu wenhua congshu (Yunnan Folk Culture Series). Lin Chaomin, ed. Kunming: Yunnan jiaoyu chubanshe.
Zukin, Sharon (1991). *Landscapes of Power: From Detroit to Disney World*. Berkeley: University of California Press.

11 Seats of Differences

Coffeehouses and the Geo-Economics of Gender in Contemporary Inner-City Tunis

Rodney W. Collins

> This problem of the human site or living space is not simply that of knowing whether there will be enough space for men in the world—a problem that is certainly quite important—but also that of knowing what relations of propinquity, what type of storage, circulation, marking, and classification of human elements should be adopted in a given situation in order to achieve a given end. Our epoch is one in which space takes for us the form of relations among sites.
>
> —Michel Foucault, "Of Other Spaces"

INTRODUCTION

From 1966 to 1968, French philosopher Michel Foucault was a purported regular at the *Café de Paris* in central Tunis, Tunisia. Foucault had accepted a teaching post in the Department of Philosophy at the University of Tunis and could easily traverse the historic medina to reach this celebrated Tunisian café. Situated at an intersection on *L'Avenue Bourguiba,* the *Café de Paris* has a storied history that emerges from the early French protectorate era as one of the first *grands cafés colonials* to be established in the *Ville Nouvelle* of the early twentieth century. The *Café* is lodged in Tunisian collective memory as one of the principal sites at which members of the *Destour* and *Neo-Destour* political independence parties met in the 1920s and 1930s. The *Café* is also associated with several waves of artistic expression. First, with a group of modernist painters known as *l'École de Tunis* in the postwar period and then during the early postindependence period, with the theatrical troupe of *Al-Firqa al-Baldiyya* and its famed director, 'Ali Ben 'Ayed, the Tunisian Othello. The *Café de Paris* is a central location where filmmakers, actors, writers, and producers attending the biannual *Journées Cinematographiques de Carthage* (Carthage Film Festival) regularly meet between film screenings. And more recently, images of the *Café* could be seen the world over as scenes of the Jasmine Revolution were broadcast through multiple channels in early 2011. Indeed, the *Café de Paris* has been situated at the emblematic crossroads of a series of critical transformations in the Tunisian sociocultural order. Consequently, the *Café de Paris* is widely considered to be a public heritage site, albeit unofficially.

At the time of Foucault's tenure in Tunis, the *Café* featured glass doors that opened onto the street, a marble bar, and a large hall interrupted with several support columns, in addition to red vinyl booths and banquettes from which Foucault may have written notes for *The Archaeology of Knowledge* or reviewed the galleys of *The Order of Things*. The *Café* may have been the vantage point for Foucault's observations of Tunisian student protests in the aftermath of the 1967 Arab-Israeli Six-Day War. The *Café* may have also provided the seat for the initiation of Foucault's homosexual relations with Tunisian students, possibly leading to his hasty departure from Tunisia.[1] Rumors that circulate concerning Foucault's tenure in Tunisia are of undeniable interest, yet it is the indexical value of the *Café de Paris* in the production of social space that is of central interest to this article. As such, this essay examines the coffeehouse as a spatial form in order to consider the dynamic imbrication of capital, gender, and space in the everyday places of the city of Tunis in the early twenty-first century.

Nevertheless, Foucault's historical patronage of the *Café de Paris* suggests some initial clues as to how a focus on the coffeehouse in Tunisia provides instructive insights into the principles that continue to organize everyday life in the city. As a gay male French intellectual, Foucault patronized a coffeehouse that provided firm and enduring footholds, carved into both the sociospatial imagination and the built environment, for each of these grossly identifying characteristics. The extent to which Foucault's identity intersects with the local sociospatial imagination in his repeated visits to the *Café de Paris* (as opposed to any number of other Tunisian coffeehouses) will be clarified through a number of ethnographic interviews that enumerate categories of coffeehouses in contemporary inner-city Tunis (i.e., *wast al-'asima*). These interviews, gathered over the course of twenty-three months between 2003 and 2008, offer multiple situated categories of the coffeehouse and lead to a consideration of the specificities of the social geographies that undergird these 'on the ground' categorizations.

Based on discussions with these interlocutors—some met by random chance and others through more tightly knit networked relations (i.e., 'snowball sampling')—I analyze the typological distribution of the range of coffeehouse institutions in contemporary Tunis. These conversations provide critical interpretations of the city of Tunis albeit in terms of coffeehouse frequentation (spatial practices), often in the form of coffeehouse classifications (representational spaces), which intimate at the role that gender and class have on the conditioning of everyday actions (spatial practices) (Lefebvre 1974; see Notar, this volume, on a discussion of the intersection between space, mapping, and cultural practices in a Chinese context).[2] Through this analysis of coffeehouse typologies, distinct zones of the city are brought into view and shown to be associated with work and leisure, men and women, tourists and locals, and above all class groupings (representations of space). These distinctions elaborate upon, and interlock with, other organizing principles (e.g., diversity, typicality) that contribute to the form of everyday itineraries in the *wast al-'asima* district.

CAFÉ À THÈME

To begin to assess the types of spatial relations that animate, and are animated by, everyday life in Tunis, I turn to an itinerary that puts into relief the diversity that emerges in the extensively regulated terrain of the coffeehouse industry. In the summer of 2008, a young architect, who went by the street name of Yassine, guided me to the leafy suburb of *Carthage-Dermech,*[3] where he wanted to show me what he considered "Tunisia's first *café à thème.*" The *Café Uranium* was located in a shopping plaza that included a bookstore and a fast-food restaurant situated directly next to a *Monoprix* grocery store; all indicators of the extent to which the 'strip mall' has successfully traversed global circuits. The plaza had at one time been promised as the future home of a multiplex theater, but this inexplicably never came to fruition. The café was expertly designed with custom Plexiglass chairs, glass countertops, lighting features, and an installation of contemporary artwork. The café certainly appeared to have an organizing design concept. However, Yassine's comments about its innovative singularity both puzzled and intrigued me, if only because Tunis—contemporary and historical—has been the site for such a wide range of coffeehouse types.

As we took a seat at a table at a less expensive coffeehouse directly across the street, I recollected an essay published in the Tunisian news magazine *Réalités* in 2004 that attested to this diversity through an investigation of the role of the coffeehouse in the southern port city of Sfax. The writers not only 'discovered' the *historical coffeehouse,* but also the *maqha al-shisha,* the *maqha al-itihad,* the *maqha al-mal wa al-'amal*[4] and one coffeehouse that was referred to locally as *maqha al-faluja*[5] due to the preponderance of discussions on conditions in Iraq within its walls (Al-Sharif 2004). One insight that I could draw from the afternoon's visit with Yassine was that of a new entry, the *café à thème*, added to the wider gamut of coffeehouse types in Tunisia: one type amongst many, rather than as an indication that the *Café Uranium* stands on one side in contrast to all other nonthematized coffeehouses on the other.

And, more critically, this young architect's statement also presented a line of inquiry to pursue: if in the context of intensive institutional codifications and regulations, the emergence of new coffeehouse forms was still possible, what might this suggest about the malleability of the sociospatial imagination? And thence, how might the diversity of these coffee-drinking institutions provide a template for apprehending the structure of difference in the city of Tunis at a time of increasing exposure to translocal flows?[6] One method of apprehending the scale of this categorical diversity and its attendant structures of difference is by sorting through the fragments of everyday conversations. I conducted semistructured interviews with Tunisian men from 'all walks of life' to gather classificatory schemas that might elaborate upon the contemporary diversity of the coffeehouse institution.

To begin, Khalil, an unmarried, male banker in his early thirties, who resides in the central district of *Bab Bhar* and generally frequents coffeehouses and bars alone, enumerated the variety of these establishments as follows: "There are four types of coffeehouses in Tunis: *café pop* or *sha'abi* which are convivial types of places; *salons de thé,* which are only in the *banlieues;*[7] *cafés d'animation/musique,* which are also in the *banlieues*; and *buvettes,* which are all over the city." Khalil's classification of coffeehouses corresponds roughly to the classification of the legal code yet also offers additional information about these categories. His classificatory schema intimates at the class dynamics that undergird his distinctions, while also indicating each institution's spatial distribution. The *café pop* or *qahwa sha'abiya* is an index of the masses, the proletariat, and the plebeian who are characterized by an open sociality. The category of the *salons de thé* suggests not only in name, but also through placement in space, an association with the upper classes and with the consolidation of capital in new urban districts. While the *café populaire* is almost invariably referred to as a *qahwa*, rare is the individual who uses the Arabic term *buqa'a shay* to refer to the newer institution of the *salon de thé*: a linguistic coding that bespeaks in a single choice phrase the impact of nearly one hundred years of French modernization efforts.[8]

Hamadi, a photographer in his mid-fifties and a long-term resident of the inner quarters of the Tunis *Médina,* suggested a different classification, one that articulates with a genealogy of social change. In the earliest moment of his reckoning, the various coffee-drinking institutions were akin to clubs that were affiliated with specific industries or regions. He offered several examples such as the *Qahwa Faluh* at *Bab Jazira* "where day laborers in construction were widely known to assemble in wait for work"; or the coffeehouse near the Administration of Television and Radio on the *Avenue de Paris*, where "not only artists but their 'agents' gathered in wait for the invitation to perform at a wedding, a circumcision or a return form the Hajj"; or the coffeehouse near *Saha Ez-Zayim,*[9] where "the masses could go, those who had less money, to find someone to play at their weddings"; or the *Café de Paris* on *Avenue Bourguiba,* "where the members of the lesser-known artistic movement, *l'École de Tunis,*"[10] gathered in the 1960s and 1970s; or the *Qahwa Univers* and *Schillinger* (pronounced: *Shill-i-ne-ger*), "where leftist journalists and students have long gathered"; or the *Qahwa al-Andulus*, with merchants on the ground floor and students upstairs; or the "*Chemantatou* where students met prior to independence."

In Hamadi's schematic, coffeehouses are associated with at least two principal factors. First, coffeehouses are associated with their most prevalent social group: artists, laborers, journalists, and students. Second, these groups reflect a relationship to the mainstream activities of the state: each group is relatively marginal to the normative functions of the social order; i.e., "leftists," "those who had less money," "day laborers," and "informal agents." The one exception to the photographer's schema is the *Qahwa al-Andulus,* where merchants and students gathered in the center of the *Médina* reflecting most fixedly a long-term centralization of the market and educational institutions at the city center.

Marwan, a thirty-year-old graphic designer/printmaker, indicated that there are both student-oriented cafés and worker-oriented cafés as he elaborated upon Hamadi's classificatory grid. Marwan segmented these coffeehouses even further by pointing out that there were those coffeehouses specifically for young male public servants referred to as *maqha 'idarat* such as the *Shilling* that might be more readily considered a *khanat*. Marwan's classificatory system also accommodated the physical organization of the city. He distinguished the *qahwa al-houma*, the neighborhood café, from the workplace café. He then classified these coffeehouses further in terms of socioeconomic factors that he exemplified as cafés in the lobbies of hotels, in *sha'abi* ('popular, working-class') neighborhoods, or the coffeehouses in the *raqi* ('refined') neighborhoods of the city. Depending upon the specific neighborhood of the city, the character of the coffeehouse could be readily assessed through the application of his criteria. Despite each of these segmentations, the designer suggested that there were also coffeehouses that were open and accessible to all people, where 'all walks of life could be found.'

A *beznassa*[11] (i.e., a young hustler) named Zied provided his short list of coffeehouse types: those that are reserved for pleasure like the *Café de Paris* and *l'Escale*, where you can drink alcohol; those that are specific to artists and actors like *l'Étoile du Nord* (*maqahi fanniya*); those that are reserved for intellectuals and professors *(maqahi 'almiya)*; those that are used by the passerby (the *flaneur*) like the *Panorama* and the *Champs-Élysées*; and, finally, those that are frequented by those who don't have work like the *Ali Baba*, also called *maqahi sha'abiya*. Zied's schema suggested a division based upon the type of social individual one is and the position one occupies in the social order; that is, his schema denoted a correspondence between an institutionalized space and the enactment of a social role.

Whereas Bassem, another young man from the streets, in his late twenties, and recently arrived to the center of Tunis from a village on the nation's western border, distinguished coffeehouses in terms of the activities that are enacted within their walls. For instance, there are the coffeehouses where he goes just to have a coffee by himself, the coffeehouses that he goes to with his girlfriend; the coffeehouses where he plays cards with his friends; and then, on a more general level, there are those coffeehouses "where you go with your best friend"; "where you go to smoke *shisha*"; the coffeehouses "where one goes if one has money" (*la bas alayhoum*), or if one doesn't (*sha'abiya* or *'arbiya*). This is a categorical order that reflects the deed, rather than the doer, and especially the structure of the social relations the doer is investing in (friendship, kinship, strangership, etc.).

Mohamed, a thirty-five-year old shoe merchant, married with children, and living in the upper-middle-class district of *Ennasr*, offered still another classificatory schema. Mohamed might be considered an expert on Tunisian coffeehouses, having confessed that he spends somewhere between four and five hours each day in a number of these institutions in various quarters of the city. According to him, the major distinctions between these coffeehouses are those that are 'popular' (*qahwa sha'abiya*) and those that are family-oriented (*qahwa 'asriya*). This distinction can then be superimposed upon a distinction between those that are located in the center of the city as opposed to those that are located out in the more refined areas of the city like *Berges du Lac*. Mohamed himself frequents coffeehouses on *l'Avenue* during the morning, often alone, as he goes about his work activities. He then frequents another type of café in the evening with friends in his neighborhood. Mohamed's schema offers two additional insights into coffeehouses and their classification: first, it provides a view into the spatial supports for the various roles that a man plays in the course of the day; and, second, it is suggestive of the relationships among urban space, gender, and capital.

Collectively, these narratives are especially instructive in terms of the sociospatial imagination that characterizes everyday itineraries as well as the values that inform those itineraries (see also Notar, this volume). These itineraries, as made evident in the views of Mohammed, Bessem, Hamadi, and Marwan, index how the city is optimally used and by whom. There is general agreement that the coffeehouses of Tunis can be evaluated in terms of several basic binaries: wealth/poverty, common/refined, work/leisure, male/female, local/foreign (i.e., tourist/passerby), and typical/atypical. These binary assessments are then interposed on the districts where they are most characteristically located; whence the *banlieues* (*La Marsa, Ennasr, Berges du Lac*) feature wealth, refinement, leisure, women (and family), the foreign and the atypical, while the popular quarters (*hayya sha'abiya*) feature poverty, the common, work (and idleness), men, the local, and the typical. These associations are stated as matters of fact whereby the assumed correlation of cleanliness, refinement, wealth, women, and the foreign becomes especially significant in discussions about the concept of *mixité*.[12]

THE GEO-ECONOMICS OF GENDER

Providing some corroboration of my interlocutors' views are the results of an ethnographic survey of the built environment of Tunis. In the early stages of my field research, I was encouraged by Tunisian intellectual Mohamed Kerrou to produce a map of the distribution of coffeehouses (see Map 11.1). This map demonstrates that of the ninety-three coffeehouses in the central district of the *Médina*, ten of these proved to be marginally *mixte*, with two of them self-described as *salons de thé*, one of which harked nominally to the famous movement *Taht Essour* and was named *Qahwa al-Sur*. Four of these coffeehouses were important historical sites on a tour of the *Médina* and included in many itineraries and guidebooks:

— *Qahwa al-Chaouachine*, situated in the *Souq al-Chaouachine*,[13] serves as both a popular spot for university students and is also heavily frequented during the month of Ramadan;

— *Café M'rabet*, claimed as one of the oldest, if not the oldest, Turkish coffeehouses in the *Médina*, is especially popular amongst young Tunisian working-class couples;
— *Café Hammouda Pacha,* an annex of a *baldi* (autochthonous Tunisian) home, which is also partitioned into a gallery and restaurant;
— *Café Rachidia,* associated with a music conservatory, which attracts young couples with a small garden.

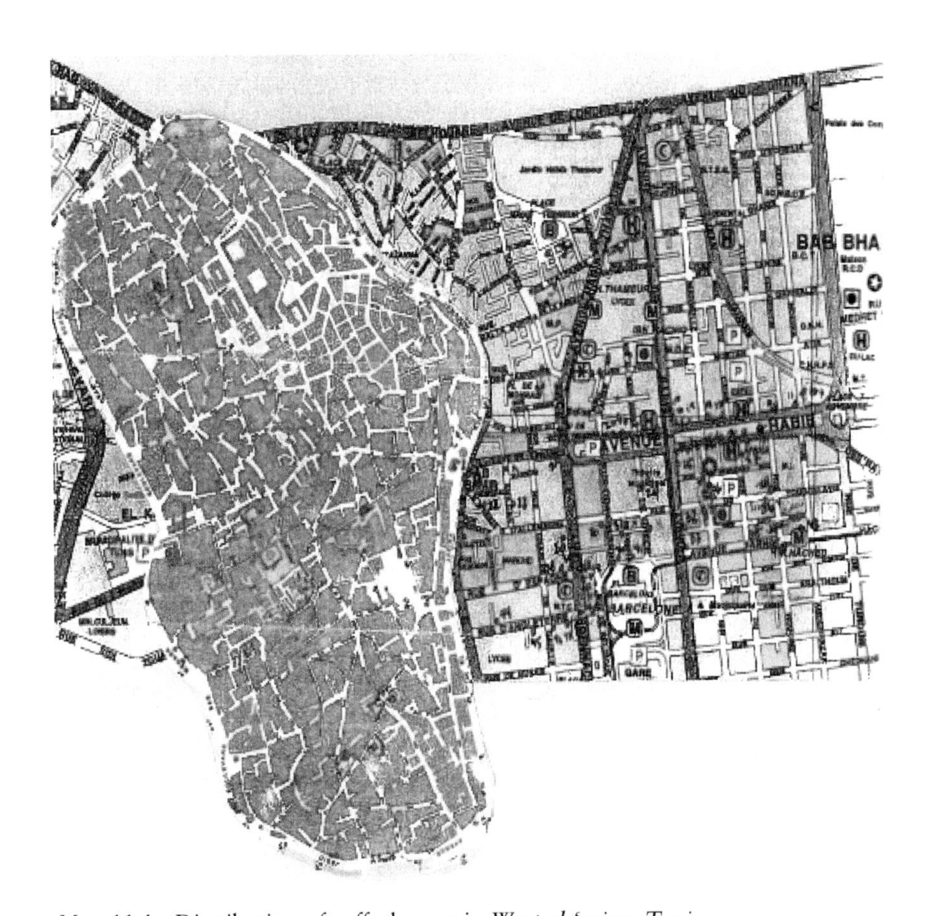

Map 11.1 Distribution of coffeehouses in *Wast al-'asima Tunis.*
The locations of the coffeehouses on the map were marked by the author (Rodney W. Collins). The map itself is constituted of a section from two different published maps produced by the author: the superimposed "medina" map is credited to the Office of Tourism, Tunisia, while the other gridded section beneath is credited to Berndtson & Berndtson, 2005 edition, Tunis City Map. Neither map in its original form indicates rights reserved.

Finally, there were five other coffeehouses that could be considered *mixte* in terms of gendered usage: a small café that was part of the *Club Tahar Haddad* cultural center, where, incidentally, Foucault offered his 1971 lecture, *Des Espaces Autres*; a small coffeehouse that was operational in the *Theâtre D'Art Ben Abdallah* as well as three *buvettes* situated in proximity to the courthouses and ministries. In each case, the coffeehouse featured a characteristic that interrupted its status as *purely* a coffeehouse, whether that feature was a music space, a theatrical venue, or a restaurant. Principally, what all of these coffeehouses had in common, aside from a mixed clientele, was that they were also mixed-use spaces.

Perspective on the spatialized division of gender in Tunis is enhanced by looking at the comparative example of the *Avenue Habib Bourguiba* ('*l'Avenue*') that lies directly perpendicular to the *Médina*.[14] Despite its proximity, a distinct gendering of space is immediately remarkable. *L'Avenue* is an approximately one-mile-long stretch of sidewalk cafés, cultural institutions, *shawarma* shops, pizzerias, and retail stores that flank a sixteen-meter-wide central pedestrian promenade lined with shade trees. The current structural features of *l'Avenue* were installed in a two-year, 2m TND renovation that was completed in 2003. Since the renovation, the number of coffeehouses and patisseries on *l'Avenue* has nearly quadrupled and the municipality of Tunis, in collaboration with the *Association de la Sauvegarde de la Médina*,[15] has implemented a unique *kiras al-shurut/cahier des charges*, or 'Code Book,' that governs the displays, façades, lighting features, and sidewalk furniture that are used on what has widely been referred to as Tunisia's *Champs-Élysées*. There are approximately thirty-seven coffee-drinking establishments along *l'Avenue*, having nearly doubled in number between the years of 2005 and 2008. Of these establishments, five were patronized strictly by male customers. That is to say, nearly all of the *cafés* and *salons de thé* on *l'Avenue* readily evidence a *mixité* of genders whereas only one in nine does so in the nearby *Médina*.

To be sure, the majority of the coffeehouses on *l'Avenue* are liberated from the official pricing categories of the government either owing to their status as *hotel-cafés*, *café-bars*, or *café-restaurants* or simply due to their location in what is officially considered a tourist zone. As such, these establishments are capable of charging two to ten times the price for a cup of coffee (i.e., 600 *millimes*–3 TND). Those coffeehouses that are only exceptionally frequented by women charge prices on the lower end of the continuum whereas those that are principally frequented by couples and tourists charge on the upper end of the continuum. Even the sale of alcohol does not fully deter the presence of women, who are regularly seen inside the *Café de Paris* or the *Café el-Hana*. Nevertheless, it is frequently the case that women who sit at the sidewalk tables of any of the establishments along *l'Avenue* are either ostensibly of high-school or university age or they are seated with children *en famille*. The most expensive of the coffeehouses attract couples as do those that have somewhat secluded outdoor seating such as *Café la Fontaine* or *Café les Pyramides*. Moreover, the practices of patronage on *l'Avenue* support the civil servant's assessment that the presence of women is dependent upon the expense of the coffee.

MIXITÉ[16]

One possible mistake to be made in assessing the landscape of coffeehouses in Tunis is to understand the painted signs that read '*café pur*' on the windows and awnings of countless coffeehouses as suggestive about the given coffeehouse's clientele (see Figure 11.1); that is, to understand the noted 'purity' as a qualification of the coffeehouse itself.[17] In fact, the signs index the type of coffee that is served in the given establishment—either pure, unadulterated coffee beans or coffee beans mixed with ground chickpeas (an additive that was used when coffee was still a costly luxury). Yet, the signs are also a useful metric for the extent of the gender mixedness or 'purity' of the clientele of these coffeehouses. That is, given a Tunisian tradition in which coffeehouses have been almost strictly patronized by men, the notion of a 'pure coffeehouse' (*café pur*) intersects broadly with the gender divisions typically associated with coffeehouses. Metonymically, these signs direct attention to the critical categorical distinction quite visible on the Tunisian public landscape between the staunchly male institutions referred to as *al-qhawi al-rijal* ('men's coffeehouses') and the

Figure 11.1 The *Café Pur.*

mixed-gendered spaces of the *cafés mixtes* (i.e., *salons de thé*). And so one might easily take the *café pur* signs as a distinction analogous to the 🚹 and 🚺 on bathroom doors that express gender directives (Lacan 1966).[18] This distinction deserves extended consideration as it articulates with a number of the divisions already enumerated by the men surveyed above, especially as it demonstrates further the embedding of gender divisions in everyday spatial practice.

For instance, when I sat down to speak with Issem, a member of the elite training forces—a *commando*—his narration of the social contours of the coffeehouse reflected a strong concern for the distinctive positions of men and women. The young *commando* lived in an apartment complex with his wife in the working-class quarter of *Zouhour*, a few kilometers away from the center of Tunis. After offering me copies of his wedding videos, our discussion turned to my research on and in coffeehouses and Issem pointed out that not only are there *cafés mixtes* and *al-qhawi al-rijal*, but that these are also found in distinct neighborhoods. He suggested that the former are found in the *centre ville* or in the wealthy *banlieues,* whereas the latter can be found in zones considered *sha'abiya*, not unlike the quarter in which he lived. Issem elaborated upon his distinction by invoking the adjective *nathif* (lit. clean, proper) that he associated with the *banlieues* in general. Further, *nathif* not only marks the difference between the coffeehouses in the *banlieues* and the *sha'abiya* zones, but also distinguishes those institutions that are reputed to be frequented by female sex workers (*binaat mahumsh bahine*) or by homosexuals (*homosexuels*).

Salim, the middle-aged director of a charitable association, assessed the presence and absence of women in coffeehouses as a question of class. He explained that the major distinction lies in the very materials from which the coffeehouse is built, including the quality of the chairs, the music, and the tables. On the one hand, he elaborated, "there are the coffeehouses of prestige, of status, and of snobbism that are mixed by principle and have only been around since the year 2000." These coffeehouses are the result not only of the influence of money, but are "the consequence of music and video clips that have expanded the range of the desire to show oneself and to be seen," from which his own teenage daughter was not immune.[19] On the other, there are the coffeehouses in the *hayya sha'abiya* ('popular neighborhoods') that are for 'everyone.' Salim, who was in his mid-forties, did not mark the contradiction that the *qahwa sha'abiya* poses: being both for the people in name and yet not for women in practice. His views suggested that the *qahwa sha'abiya* is an institution that naturally suits a categorical understanding of 'the people' that precludes and, at least in the context of the practical normativity of Tunisia, predates the full participation of women (see Perry's discussion of gendered space and political activism in Salvador, Brazil, this volume).

This segregation of gendered space is especially striking in light of the legacy of reform surrounding women's rights in Tunisia. In the course of the 1930s *Taht Essour* literary movement, Tahar Haddad emerged as a front-runner in the women's emancipation movement. His writings were critical to the position adopted by President Habib Bourguiba and effected in the nation's landmark *Personal Status Code* signed into law in 1957. The code abolished the practice of polygamy, gave women the right to divorce their husbands, and fixed the legal age of marriage at seventeen. Subsequent Tunisian legislation endowed women with the right to a first-trimester abortion in 1965, undoing the French Napoleonic criminal code even before similar legislation in France in 1975. The availability of the abortifacient RU-486 (*Mifepristone*) in Tunisian pharmacies was nearly coincident with that medication's availability on the American market. Moreover, by 2008, nearly one-third of Tunisian judges and lawyers were women, as were a quarter of elected parliamentary delegates.[20] Despite these enormous strides in women's juridical rights, everyday practice suggests a rigidly entrenched gender division with one symptom being the distribution of bodies in Tunisian coffeehouses as well as across the urban landscape. One female academic who spoke with me about the nature of the Tunisian coffeehouse expressed her regret that there was not such an institution that she could make use of. "Yes," she admitted, "there are the *salons de thé* in the *banlieues*, and there are the hotel cafés but what I would really prefer is one of those neighborhood *qhawi* that men are able to make use of and where the cup of coffee is only 300 *millimes*."[21]

This gendered division of space (and spatialized division of gender) does not rely on any official legislative policy that explicitly prevents women's presence in those coffeehouses considered *al-qhawi al-rijal*. Rather, quite paradoxically, the coffeehouses most constrained by state economic policies are the same coffeehouses that evidence least the progressive gender politics championed by the state. A public servant in his mid-thirties, who specializes in the accountability of government agricultural projects, explained to me that the contemporary gender division in coffeehouses is a simple one to understand:

> Things changed 10 years ago, that's when women really began to have their own salaries and their independence. They needed somewhere different to spend their time than men. This is why there is the phenomenon of salons de thé. Over time it is possible that the quality of the salons de thé, which are always more expensive, and the regular coffeehouses for men will approximate one another but only in the sense that the men's coffeehouses will receive a 'lifting' of quality. But you've got to remember that men are also changing. Although, you'll never see two men sitting in a salon de thé, you'll only ever see two women, or a woman and a man. I suppose, over time, you will see a slow disappearance of these men's coffeehouses, but very gradually and never totally (June 6, 2008).

In addition to an explicit explanation and demarcation of coffeehouse categories, the public servant's account is implicitly characterized by two specific anxieties. The first anxiety is prognostic and is related to his assessment of a transformation in men's subjectivities that can be understood when evidenced in representational spaces and spatial practices. The limit of that transformation as indicated in two men seated in the 'comfort' of a *salon de thé* attests to the tenuous, though persistent, delineation between the constructs of gender. This is confirmed by the instructive irony that is expressed in the public servant's assessment that eventually these *qhawi al-rijal* will receive a 'lifting' of quality as influenced by the *salons de thé*. The second anxiety is diagnostic and relates to the radical status accorded to women in Tunisia, especially when viewed in comparison with its Arabo-Muslim neighbors. This comparative edge is widely attributed to the initiatives of the aforementioned visionaries Tahar Haddad and Habib Bourguiba.

STRIDES TOWARD GENDER NEUTRALITY

At least one recent effort has been made to interrupt this imbrication of space, gender, and capital in central Tunis. Nayla, a self-described militant Tunisian feminist, recounted how she was part of an initiative to take back the coffeehouses as public spaces in the early 1990s. Meeting at a fast-food chain in the affluent suburb of *La Marsa*, Nayla told me how each day for several weeks, she and ten to fifteen of her female colleagues, associates, and friends gathered in the *Café de Paris* on *l'Avenue*. Their daily meeting time was set for 4 p.m. During their meetings, they discussed issues related to the role of women and their place in the social life of Tunis. Their objective was to do nothing more than occupy public space; this was the task in and of itself. The *Café de Paris* was undergoing renovations and the women capitalized on the possibility of reclaiming the *Café de Paris* for a female clientele. Their arrival at the coffeehouse was well received and the servers were professional and friendly. Other clients in the coffeehouse did not take much notice of the large group of women amongst them, including some who now work in important civil-servant roles. It was a matter of "acting politically and publicly without entering into the mechanisms of the state as with so many of the contemporary women's groups." Nayla concluded that "the initiative, unfortunately, went almost entirely unnoticed."[22]

Despite her participation in this activist initiative, Nayla admitted that she does not feel entirely comfortable in other coffeehouses on *l'Avenue* such as *l'Univers*. Rather, her comfort is found in the coffeehouses near the *Champion* supermarket in *Lafayette*, in modern areas ("*dans des zones modernes*"). Nayla's position was instantiated in our meeting at an outlet of the *Baguette Baguette* sandwich franchise. This sense of comfort is attenuated by Nayla's sense "that":

> The strides of women in Tunisia—at their best in the age of Bourguiba especially during the 1970s—are smaller and smaller with each passing year. Women have become influenced by their visits to the Middle East where they have worked as teachers for several years, returning to Tunis with conservative ideals, influenced by their husbands and brothers who have also traveled.

Nayla suggested that women frequent less and less the public coffeehouses—not only because of the encroachment of conservative values from the Middle East, but also due to new habits of visitation whereby the *petit bourgeois* invite their friends and associates to their large, modern homes.[23] This frustrates Nayla's view of the coffeehouse that she praised as "a neutral space, a space for encounters where one is not 'censored' by the intimacy and the politeness that comes with a visit to someone's home and which precludes criticality of thought." She contended that "the younger generation does not see what they are leaving behind, what they have forfeited which extends to even the women who no longer leave their offices to eat lunch." This, she argued, has only been reinforced by "the shoddy social science research that enters into publication by women like D. M. who simply polish up outdated studies which no longer reflect the reality of Tunisian women, but reflect a golden age that has long come to pass."

Nayla's comments and reflections underline the perceived potentiality of the coffeehouse as a neutral space that influences the psycho-philosophical disposition of the individual. Neutrality is also a theme suggested in prominent theoretical approaches to the coffeehouse (Habermas 1989; Sennett 1992). Yet, Nayla's narrative, in negative relief, also provides a mapping of the urban landscape that coincides more closely with the regime of classifications accounted for by Tunisian men. Those classifications suggest that the coffeehouse is rarely a zone of neutrality because it is built upon conditions that regulate the mixing of male and female bodies. This constraint is fortified by what are suggested to be the gendered costs of patronage. The prices of coffee-based beverages are controlled and fixed by the state in all *1ère catégorie DBOs*. Exceptions to this pricing structure are made for a certain number of *salons de thé* that may justify their divergence from the state-pricing schema by way of investments in décor and furnishings.

SEATS OF DIFFERENCE

Through a series of images of coffeehouse chairs, I aim to provide a glimpse of the type of difference that is facilitated by capital investments in contemporary coffeehouses, as well as the materialities of gendered use. First, it should be borne in mind that Article 21 of the *cahier de charges* insists upon the presence of chairs and tables adequate for the size of a coffeehouse and its clientele, thereby distinguishing it from a *buvette,* which, in contrast, is forbidden to provide chairs for its clientele. As such, chairs are a feature that specify not only what kind of *DBO* one is in, simply due to their absence or presence, but they also provide a richer semiotic for the expression of gendered divisions. The chairs in the first two photos were found in a coffeehouse in the neighborhood of *Bab Bnat*—a residential quarter that flanks the west side of the *Médina*. The *qahwa* pictured in these two photographs is named *Saadi* and is frequented expressly by male customers of various ages (see Figures 11.2 and 11.3). The chairs are stackable, easy to move and clean, without padding, and of low cost.

Figure 11.2 Chairs at *Qahwa Sa'adi.*

Figure 11.3 Chairs at *Qahwa Sa'adi.*

The padded chairs in the third photograph are part of the furnishings in the *Grand Café du Théâtre Salon de Thé* on *l'Avenue* frequented by men, couples, and the occasional duo or trio of women. The chairs are only minimally padded and made of varnished wood (see Figure 11.4). Although their distinctiveness from the chairs at *Saadi* is minimal, they provide a material semiotic fashioned of foam and fabric that is understood by both men and women who regularly make use of them.

A final example demonstrates more fully this material relationship between space, gender, and economics. The chairs in the final image are part of the furnishings in a *salon de thé* in the district of *Ennasr* named *Touareg* (see Figure 11.5). The chairs are especially remarkable given their heaviness and thick padding, almost as though furniture meant for a private home. An interview with the owner of the *Touareg* provided details on the extent to which he had gone to engender a clientele of primarily mixed couples, mixed groups as well as duos and trios of women.

Figure 11.4 Chairs at *Grand Café du Théâtre.*

Figure 11.5 Chairs at *Salon du Thé Touareg.*[24]

> Our theme is the desert. The designer thought that people feel most at ease in a desert setting, relieved, transported. If you know your history, you'll know that the Touareg are a tribal people that live in the deserts, in the southern part of the Morocco, although not in Algeria or Tunisia. You'll also know that there is a Volkswagen, a 4x4, called the *Touareg.* So the designer chose this theme, he's really a professor of fine arts, a sculptor, not just a laborer. Not a worker. He chose the theme and he began to build it like the southern city of Matmata, like a house in that part of the country. He's a real artist. Notice even the wood is palm wood. So, every detail has to support this theme, the music is on at a certain volume, not too loud so we don't get too many young kids. We like soft music, and yes, we've got a plasma television but we mostly use it to play video clips. And there are no *shishas* because that takes too much time whereas a coffee will take at most an hour. This place is one of those places you come to because that is all you want to do; you don't stop here while you are in the markets, shopping. You come here as a destination (June 12, 2006)

In line with the public servant's assessment of the relationship between physical comfort and gender, as well as Nayla's increased sense of psychological comfort in what she referred to as more modern establishments, the *Touareg* was frequented primarily by duos and trios of women. Indeed, the owner of this *salon de thé* ambitiously pursued his clientele through his choice of furnishings.

Interestingly, the privileging of women in this case, which attests to an association of female space with refinement, or with a 'lifting,' of quality, offers a double inversion of the classic "male is to culture as female is to nature" principle. First, the woman is associated with the refinement of furnishings (padded chairs, fine woods, original art) that places her in a relationship with culture (Ortner 1974, 1996). Then, it is made evident that the space is organized according to a natural schema: the desert, thereby reinserting the woman in a proximal relationship with nature. Next, the space is referred to as not being like any other public space, but 'like a house,' thereby reasserting the association of women with the domestic space, and reinforcing her perceived distance from the cultural realm. The fractal recursivity of the "male : culture :: female : nature" principle reaches its limit, however, in one type of client who very much evaded observation: the lone woman. While the circulation of women in diverse configurations is very clearly supported by the structural amenities of the *salon de thé*, it seems that the social order promotes an invisible value that deters the appearance of the lone woman, 'the hidden space of the sacred' (Foucault 1986). Indeed, the presence of a single woman in a *salon de thé* might just well indicate a failure, rather than an emancipatory success, of the Tunisian social system in which such a woman moves.

THE UTOPIA OF THE HETEROTOPIC

There is, however, something markedly different about the *Touareg Salon de Thé* and the site of Nayla and her comrades' actions. As the coffeehouse of choice for their initiative, *Café de Paris* emerges once more as something of an exception, a café which suggests itself as the spatial equivalent to the open plaza of other cities where protest and politics forcibly emerge (Low 2000). To be sure, contemporary Tunisia arguably has had few formidable spaces of public protest and thus the café is especially significant. The *Café de Paris* operates under the *3ème* administrative category which enables service of *Ricard* and other distilled spirits, in addition to beer and coffee. It is precisely the type of café that would have been outside of the official purview of indigenous Muslims during the Protectorate period, serving instead to fortify the *metropolitan* social order in the city of Tunis. Its subsequent and alleged use as a meeting spot for protonationalists in the 1930s has only served to embolden its legendary status (Belaid 2004).

To hear Nayla speak of the coffeehouse as a zone for public action in the early 1990s reiterates its perceived potentiality as a site from which social change could be performed. Further, in the summer of 2006, a group of young Tunisian men engaged in a campaign for HIV/STI prevention used the café as a site from which to distribute condoms and to raise social awareness around the dangers of unprotected sex. It is also the site where tourists come with young men who are engaged in the practice of *bezness*—a form of hustling that enables shifts across socioeconomic barriers. Indeed, the *Café de Paris* holds promise as a ground for emancipatory or transgressive action: whether the terms of that action are class, race, gender, sexual orientation, or nation based.[25] The café continues to function in the popular imagination as a locus for the performance of resistance and difference, thereby retaining and perpetuating the long-term association of the coffeehouse with counterhegemonic activities.[26] It is probable that due to the absentee corporate ownership of *Café de Paris*—today's *Café de Paris* is owned and operated by the national beverage company *Société Frigorifique et Brasserie de Tunis* (SFBT)—divergent activities are 'permitted' within its walls; the association of illicit, illegal, or seditious actions in the space of a coffeehouse owned by an individual owner would conceivably carry some risk for the owner, especially in a state characterized as a police state.

Interestingly enough, Michel Foucault penned his well-known lecture *Of Other Spaces, Heterotopias* in 1967, when he was resident in Tunis. The lecture, published in 1984, indexes his exposure to the 'Orient'[27] and is arguably reflective of his practices in visiting the *Café de Paris,* a *heterotopia* of quite extraordinary potential (Augé 1992).[28] Indeed, the *Café de Paris* operates as a heterotopia of deviation whereby at the very center of the city of Tunis is situated this coffeehouse that occasions the non-normative behaviors for which Foucault himself may have been suspected. The *Café de Paris* with its wide open doors and large hall appears as an open and accessible public space, but also instructively exposes the exclusions that are everywhere in the city of Tunis. For Nayla, the *Café de Paris* operated as a *heterotopia* of compensation wherein the promise of Habermasian neutrality was to be had and where 'all walks of life' were welcome. For the ranks of the unemployed in today's neoliberal Tunis, like the young *beznassa* interviewed in the course of this article, the *Café* serves as a haven for a deviant idleness, thereby obviating the social differences that otherwise riddle the material and ideological firmament of the city.

CONCLUDING THOUGHTS

This chapter began with a historical profile of the famed Tunisian *Café de Paris* in order to set the stage to examine the mutually dynamic imbrication of gender and capital in the built environment of contemporary Tunis. I argue that the coffeehouse form provides a means through which a Tunisian sociospatial imagination and its attendant structures of difference can be assessed. This assessment is leveraged on the basis of observations and narratives of coffeehouse patronage (spatial practices) and coffeehouse typologies (representational spaces). By evaluating the typological range of coffeehouse forms, I demonstrated the manner in which the city is segmented in the everyday classificatory schemas of several Tunisian men and one activist woman. Although several of these interlocutors suggested the neutrality of the coffeehouse, indicating that coffeehouses were suited for 'all walks of life,' few specific examples were forthcoming. Instead, coffeehouses were shown to be ideologically and materially oriented to specific clientele: trade, status, class, sexuality, and gender were indicated by a number of my interlocutors as major organizing principles. And, despite the resultant diversity, a sense of typicality persists and is especially evident in entrenched gender and social divisions.

The most oft-remarked typological division was that between *cafés mixtes* and *al-qhawi al-rijal*, a gendered division that was further evidenced in terms of coffeehouse distribution in the urban landscape, as well as reinforced through the material specificities of furnishings and décor. Consideration of the concept of *mixité* led to an examination of the heterotopia of the *Café de Paris*, a coffeehouse that evidently interrupts, by competitively exposing, the recalcitrant spatial divisions of everyday life in contemporary Tunis. To demonstrate how recalcitrant these divisions are, I introduced the case of a Tunisian activist who strode to interrupt them. Yet, even in Nayla's case and that of the *Café de Paris*, it is evident that movement through the social and physical space of Tunis is shaped quite extensively by the logic of capital flows: where capital investment is at its densest, women are most visible, whereas broader class-based exclusions are attenuated in the process. Consequently, an analysis of the position of the coffeehouse in the sociospatial imagination demonstrates that structures of difference in Tunis can be effectively mapped by tracking the flows and deposits of capital in the city whereby particular types of gender *mixité* are instantiated through the maintenance and proliferation of coffeehouse forms.

ACKNOWLEDGMENTS

The ethnographic research that forms the basis of this essay was made possible by grants from the Wenner-Gren Foundation for Anthropological Research, the Fulbright Foundation, and the Columbia University Middle East Institute. Writing support was provided by a Ford Foundation Diversity Fellowship and the Georgetown University's Center for Contemporary Arab Studies Qatar Postdoctoral Fellowship.

NOTES

1. Edward Said is the source of this allegation:

    > Although we chatted together amiably it wasn't until much later (in fact almost a decade after his death in 1984) that I got some idea why he had been so unwilling to say anything to me about Middle Eastern politics. In their biographies, both Didier Eribon and James Miller reveal that in 1967 he had been teaching in Tunisia and had left the country in some haste, shortly after the June War. Foucault had said at the time that the reason he left had been his horror at the 'anti-semitic', anti-Israel riots of the time, common in every Arab city after the great Arab defeat. A Tunisian colleague of his in the University of Tunis philosophy department told me a different story in the early 1990s: Foucault, she said, had been deported because of his homosexual activities with young students. I still have no idea which version is correct.

    See http://foucaultblog.wordpress.com/2008/06/21/said-on-foucault/. Said's suspicion seems somewhat misplaced given that Foucault returned to Tunis only two years later to offer lectures of his work as well as sit for interviews with Tunisian journalists. See: "Folie et civilization" lecture excerpted in La Presse April 10, 1987. The original date of the lecture was April 24, 1971 at Club Tahar Haddad. For the full lecture and a discussion of Foucault's presence in Tunisia, see Triki, R., 1989. See also the interview with Hafsia, J., 1971.
2. My considerations of spatiality and spatialization follow the basic premise of Henri Lefebvre's argument that space is a set of relations that are produced at the intersection of *representations of space, representational spaces* and *spatialized practices.*
3. Carthage-Dermech may also be familiar to the reader as the alleged site of the second century B.C.E. Punic sacrifices of children that remain of controversy to archaeologists and historians of the ancient settlements of Carthage.
4. "Shisha café," "Labor Union café," and "Money and Work café."
5. "Al-Falluja café."
6. Officially, there are only three categories of coffee-drinking institutions in Tunisia, that is, *débits de boissons* ('beverage vendors') belonging to the *1ère, 2ème* and *3ème* categories. In 1898, the colonial authorities of the Tunisian protectorate (1881–1956) established these three categories principally as a means of delimiting the sale of alcohol, a policy that dovetailed with

an interest in preventing its sale to indigenous Muslims. In everyday parlance, there is an even larger range of categories into which the estimated 18,000–27,000 of these establishments are segmented, most of which are not apprehended by the state's classificatory schema of *débits de boissons*. For a historical approach to a number of these categories by a Tunisian documentary filmmaker, see: Ladjimi 2001.

7. I should make clear, given the political charge of the term in French contexts, that the *banlieues* that are referenced herein are the residential and commercial neighborhoods of the wealthiest members of Tunisian society.
8. Tunisia declared Arabic to be its national language at the time of independence. Nevertheless, official documents were printed in both French and Arabic until 2001. So profound has been the linguistic imprint of the French colonial order that even forty years after independence, the Tunisian Ministry of Education continues to be engaged with the Arabization of the national educational curriculum. For an informative discussion on the Arabization of the instruction of mathematics, see Abd Allaoui, R., 1995. For a discussion of the related concepts of *surri* and *'arbi*, see Zussman, M., 1992.
9. 'Leader's Square.'
10. The "École de Tunis" was an artistic movement that was initiated as early as 1948, but which found its greatest expression in the years postindependence. Members of this group include Yahia Turki, Abdelaziz Gorgi, and Jellal Ben Abdellah.
11. The practice of *bezness* and the subject position of the *beznassa* are especially complex and I elaborate on these themes in Collins, R., 2009. For the purposes of this article, I allow for the rough translation of 'hustler.'
12. An additional classification of coffeehouses that was not mentioned by any of these interlocutors is the 'foot-café.' My interlocutor's failure to mention this class of coffeehouse was likely due to the fact that all coffeehouses with televisions at the time of major football matches operate as 'foot-cafés.' This will be discussed further below. For a journalistic approach to the 'foot-café,' see Almi, H., 2008.
13. For an early anthropological study of the traditional Tunisian red hat known as the *chéchia*, see Ferchiou, S., 1973.
14. For an instructive study on the relationship between gender and space in Tunisia, see Zannad, 1984.
15. 'Association for the Protection of the *Médina*.' 24 rue du Tribunal, Tunis, Tunisia.
16. When discussion of mixed gender spaces and practices in Tunisia are denoted in French, this is accomplished with either the noun *mixité* or with the adjective *mixte*. In Arabic, the word *mukhalita* is generally used.
17. In French, the term 'café' is used for both the beverage and the locale, thus providing the grounds for this ambiguity.
18. Cf. "Ladies and Gentlemen will be henceforth two countries towards which each of their souls will strive on divergent wings, and between which a truce will be the more impossible since they are actually the same country and neither can compromise on its own superiority without detracting from the glory of the other." Lacan, J., 1977, page 152.
19. For an early, though intriguing, anthropological discussion on being 'seen' in the Islamic City, see Gilsenan, M., 1982.
20. Source: *Centre de recherches, d'études, de documentation, et d'information sur la Femme*, Avenue du Roi Abdelaziz Al Sâoud, Rue Farhat Ben Affia, El Manar II–2092—Tunis, Tunisia.
21. The Tunisian *dinar* (TND) is constituted of 1,000 *millimes;* at the time of the interview in 2008, 1 TND = .75 USD.

22. There is one remark on this initiative in a document published by the *Federation Nationale des Villes Tunisiennes* (*FNVT* 2001). Furthermore, it is questionable to what extent this initiative went unnoticed. My best interpretation of this comment is that the initiative went unnoticed by other women who did not join the initiative. This interpretation is in line with Nayla's continued disappointment with the activities of her female contemporaries.
23. For a rich, ethnographic account of women's visitation practices in urban Tunis, see Holmes-Eber, P., 2002.
24. This photograph was taken after these statements from the *salon de thé*'s owner: "How many photos do you want to take? One, or, a lot? I'll tell you why I ask: you see, there's a lot of piracy in Tunisia and if I come up with a good idea, and then I let you photograph it and then you do the same thing, then I will lose. I usually don't let people take photographs, but you can go ahead and take a couple."
25. The relationship between the official classificatory status of the *Café de Paris* to its practical, imagined, and historical use by individuals in Tunisia reiterates the political effort around which much social science work pivots, a thesis which is quite extensively treated in the work of James C. Scott's (1998) *Seeing Like a State,* although more pithily encapsulated by Alfred Korzybski's dictum: *the map is not the territory.* This is, in fact, the case with the majority of Tunisian coffeehouses. Although these establishments are heavily codified, zoned, and policed, they provide a wide gamut of diversity for their everyday clientele and their roles in individual narratives.
26. At a time of increased strains on the average Tunisian's budget, *SFBT* posted a 7 percent increase in net revenues during the first six months of 2008.
27. Note Foucault's usage of the examples of the 'Oriental garden,' the Persian rug, the 'Moslem hammam,' the 'Polynesian huts of Djerba.'
28. The concept of the *heterotopia* is modified and provided ethnographic treatment in Augé, M., 1992.

REFERENCES

Abd Allaoui, R. (1995). *"Al-'arabiya luga . . . wa al-latiniya ramuz," in* Réalités 518 (6), October 14–15.

Almi, H. (2008). "Foot-cafés!" in *Le Temps* 6 February; last accessed November 6, 2009, http://www.letemps.com.tn/pop_article.php?ID_art=12898.

Al-Sharif, M. (2004). *"Bayna al-shisha wa al-niqaba wa al-siyasa,"* 6–12 May, *Réalités* 958: 12–14.

Augé, Marc (1992). *Non-lieux: introduction á une anthropologie de la surmodernité.* Paris: Seuil.

Belaid, Habib. (2004). *"Le Café Maure en Tunisie à l'époque colonial." Arab Review for Ottoman Studies* 29: 47.

Collins, Rodney (2009). From Coffee to Manhood: Grounds for Exchange in the Tunisian Coffeehouse, ca. 1898–2008. PhD dissertation, Department of Anthropology, Columbia University, New York.

Ferchiou, Sophie (1973). *Technique et Société, la fabrication de la chéchia de Tunisie.* Paris: Clermont Ferrand.

FNVT (2001). Tunis City Development Strategy Report *Federation Nationale des Villes Tunisiennes (CDS, Tunis)* pgu.tunisie@planet.tn.

Foucault, Michel (1986). "Of Other Spaces (Des Espaces Autres)." *Diacritics* 16: 22–27.

Gilsenan, Michael (1982). *Recognizing Islam: Religion and Society in the Modern Arab World.* New York: Pantheon Books.
Habermas, Jurgen (1989). *The Structural Transformation of the Public Sphere: An Inquiry into a Category of Bourgeois Society.* Cambridge, MA: MIT Press.
Hafsia, J. (1971). "Un problème m'interesse depuis longtemps, c'est celue du système penal." *La presse de tunisie* August 12: 3.
Holmes-Eber, Paula (2002). *Daughters of Tunis: Women, Family and Networks in a Muslim City.* Boulder, CO: Westview Press.
Lacan, Jacques (1977) [1966]. *Écrits.* New York: Norton.
Ladjimi, Mokhtar. (2001). L'Orient des cafés *Paris: Casadei Productions.*
Lefebvre, Henri (1991) [1974]. *The Production of Space.* Cambridge: Blackwell.
Low, Setha M. (2000). *On the Plaza: The Politics of Public Space and Culture.* Austin: University of Texas Press.
Ortner, Sherry B. (1974). "*Is Female to Male as Nature Is to Culture?*" *Woman, Culture and Society,* ed. by Michelle Rosaldo and Louise Lamphere. Stanford, CA: Stanford University Press: 68–87.
Ortner, Sherry B. (1996). *Making Gender: The Politics and Erotics of Culture.* Boston: Beacon Press.
Scott, James C. (1998). *Seeing Like a State: How Certain Schemes to Improve the Human Condition Have Failed.* New Haven, CT: Yale University Press.
Sennett, Richard (1992). *The Fall of Public Man.* New York: W.W. Norton.
Triki, R. (1989). "Folie et civilization. Foucault en Tunisie." *Cahiers de Tunisie* 39.
Zannad, Traki (1974). *Symboliques Corporelles et Espaces Musulmans.* Tunis: Cérès Productions.
Zussman, Mira (1992). *Development and Disenchantment in Rural Tunisia: The Bourguiba Years. Boulder: Westview Press.*

12 From the "Margins of the Margins" in Brazil

Black Women Confront the Racial Logic of Spatial Exclusion[1]

Keisha-Khan Y. Perry

> The right to the city is like a cry and a demand. This right slowly meanders through the surprising detours of nostalgia and tourism, the return to the heart of the traditional city, and the call of existent or recently developed centralities . . . The *right to the city* cannot be conceived of as a single visiting right or as a return to traditional cities. It can only be formulated as a transformed and renewed *right to urban life*. (Lefebvre, Kofman, and Lebas 1996: 158)

> Brazil as a nation proclaims herself the only racial democracy in the world, and much of the world views and accepts her as such. But an examination of the historical development of my country reveals the true nature of her social cultural, political and economic autonomy: it is essentially racist and vitally threatening to Black people. (Nascimento 1989: 185)

SOCIAL PROTEST FROM THE MARGINS

On March 20, 1997, women in Gamboa de Baixo prepared in the darkness and silence of early morning. Late the night before, residents had received the shocking news that fourteen-year-old student Cristiane Conceição Santos had died from head injuries. The fatal car accident on her way to school was one of three violent incidents at the beginning of that year alone involving Gamboa residents crossing one of Salvador's busiest streets, Avenida Lafaiete Coutinho (Contorno Avenue). In those accidents, one person had died and another became paralyzed. Ironically, the week before, women from their neighborhood association, Associação Amigos de Gegê Dos Moradores da Gamboa de Baixo, had attended a meeting with the mayor during which they requested that he install traffic lights and a crosswalk. They had insisted on the control of traffic for the safety of pedestrians, including women and young children, who risked their lives on a daily basis just to attend school or to go to work. "No one respects those of us who try to cross the street," one woman told a newspaper reporter during the protest. The death of Cristiane was a brutal reminder that they had received no official assurance that safety conditions on the road would change.

As dawn approached, these women hurried together with their children into the street, determined to walk without fear on the Contorno Avenue that passed above their community. As Gamboa de Baixo activist Maria remembers, "Before [the closing of the street] we fought with fear, but the day we closed the Contorno [Avenue], full of courage to confront the police, I felt that I had a space in this society that's mine" (personal interview 2000). Anger intensified as the sun rose. They moved quickly to get the word out to other neighborhoods, black movement activists, NGOs, and supporters in the local archdiocese. Between 6 a.m. and 8 a.m., the Contorno Avenue was the site of one of Gamboa de Baixo's and the city of Salvador's largest and most significant acts of public defiance of the late 1990s. Residents held a banner that stated, "Gamboa de Baixo Coveted Paradise (Paraíso Cobiçado) Demands Help." The demonstration disrupted the normal flow of traffic throughout the entire city. They blocked the street with burning tires, wood, and other debris. The main actors in this manifestation were black women, young and old, who shouted in defense of their families and their communities. The fire department extinguished the fires and removed the debris to open the congested Contorno Avenue to transit. The military police (in their riot gear) stayed the entire morning to prevent the outbreak of new demonstrations.

For the black women activists of Gamboa de Baixo, the Contorno Avenue protest I describe above represents one early memory of the neighborhood's grassroots struggle for permanence, land rights, and social and economic change in the area. Gamboa de Baixo is a predominantly black century-old fishing community located on the coast of the Bay of All Saints (Baía de Todos os Santos) in the center of Brazil's city of Salvador in the northeastern state of Bahia. I first heard this story described by local activists when I began this ethnographic research on how black women lead social movements in 1998 and 1999, while living in Salvador and marching alongside these black women in the celebration of the *Dia Nacional da Conciência Negra* (National Day of Black Consciousness) on November 20 in the center of Salvador. During the past decade, as I continued the research, neighborhood activists have participated in several of these types of protests on the Contorno and throughout the city, some of which are promoted by larger black organizations such as the United Black Movement (Movimento Negro Unificado) and the Black Union for Equality (União de Negros pela Igualdade, UNegro). Public demonstrations, oftentimes spontaneous and disruptive to the urban social order, have focused on black concerns with increased police violence and the unemployment and poverty that disproportionately occurred in their communities. Gamboa de Baixo street protests have been a way for poor black people to claim power and space during times when urban-renewal projects are forcibly removing them from these central areas of the city. In fact, I argue in this chapter that Gamboa de Baixo's participation in the November 20 events every year has been an expression of a shared understanding of the linkage between the social conditions of poor blacks and structural racism in Bahia and throughout Brazil. I assert that social actions organized by black women at the neighborhood level should be understood as the everyday political manifestations of the broader Brazilian black movement. (Notar and Collins, in this volume, also discuss, although in different ways, the complex intersections between urban change, class, and gender in Kunming, China, and Tunis, Tunisia, respectively.)

This chapter historicizes the Gamboa de Baixo community movement against urban renewal programs and for access to material resources such as urban land and housing (on the issue of urban renewal, see the chapters by Zukin, Forrant, and Ryan in this volume). Collective memory of how these urban renewal projects have rapidly transformed Salvador and the everyday lives of black neighborhoods is significant for understanding the political formation of black women throughout the city. This movement emerged because, as Gamboa de Baixo residents understood it, they were most certain to become the next Pelourinho, Salvador's Historic Center (Centro Histórico). The government removed the poor black population from the Pelourinho during that neighborhood's "revitalization" process during which Gamboa de Baixo political organizers wrote in the neighborhood communiqué in 1995: "We do not want to be a second edition of the Pelourinho." Almost fifteen years later, in 2010, despite more than a decade of struggle and significant infrastructural improvements to the neighborhood, the threat of expulsion remains at the center of Gamboa de Baixo's political organization. The 2009 displacement of Vila Brandão, a coastal black community located between the upper-class neighborhoods of Barra, Graça, and Vitória, continues to galvanize the women activists in Gamboa de Baixo to organize against the very real possibility of expulsion. This black women-led grassroots struggle is even more urgent since the August 2010 government demolition and "cleanup" of the traditional beach cabanas (*barracas*) vividly illustrate. Many Gamboa de Baixo residents express that it is only a matter of time before the bulldozers will also come to clear their coastal lands. The case of Gamboa de Baixo, specifically, and coastal land expulsions, more generally, demonstrate the particularities of Brazilian antiblack racism and black political organization against institutional racism. Through my analysis of their narratives and methods of resistance, I argue that discourses and practices of urban renewal are prime examples of antiblack racism in Brazil, that political mobilization against urban renewal in Salvador illustrates the racial consciousness of Brazilian blacks, and that black communities *do* actively organize around race and gender in political struggles over material resources. To support these claims, I focus on the interrelationship between racial and gender-identity politics and struggles over resources, particularly housing and land, in the city center of Salvador.

In order to frame the urgency of the black women–led urban mobilization of Gamboa de Baixo residents within the discourses and practices of the broader black movement in Bahia and Brazil, I organize this chapter into three sections. First, I outline the spatial and political context for the emergence of the Gamboa de Baixo neighborhood with a focus on the racial and class logic of urban renewal. Second, I examine the place of history in commodifying black communities while simultaneously excluding them. Finally, I conclude with the summation of the significance that urban social movements have on the formation of the black movement in Brazil. This analysis draws primarily upon documents and activists' narratives produced in the late 1990s and early 2000s at the height of the neighborhood movement against expulsion.

"OUT OF PLACE": CHANGING THE FACE (RACE) OF THE CITY CENTER

In Bahian tourism politics, both the city center and all areas along the shore of the Bay of All Saints are strategically important for the development of leisure and cultural sites (Fagence 1995). During the past two decades, Salvador's municipal government has implemented a series of projects intended to recuperate, restore, and "revitalize" the environment of the urban center and along the coast. Founded in 1949, the city was Brazil's first capital and still holds some of the country's most historically significant monuments and buildings. In 1981, UNESCO recognized Salvador's Historic Center as a World Heritage Site (Collins 2008; Dunn 1994; Pinho 2010; Romo 2010; Sansi 2007). Government officials forcibly removed the poor black population from the Pelourinho that they considered "dangerous" and "criminal," and that occupied some of Brazil's oldest colonial homes. Since the 1990s, the city has spent millions of dollars on revitalization projects, such as the Bahia Municipal Development Project, which have restored the historic buildings of the city center such as churches, forts, and mansions. The Pelourinho now represents one of the most important landmarks visited by national and international tourists every day.

In the recent memory of Salvador, the gentrification of the Pelourinho is a symbolic marker of black experiences with mass displacement and the repressive regimes of urban restructuring (Pinho 1999). The restoration of the Pelourinho marks the beginning of urban-renewal programs that have usurped the city center and other parts of Salvador during the past two decades. Whereas the government has subsequently restored historic monuments such as the São Marcelo Fort, the Pelourinho is significant because of the expulsion of the local population and its violent relocation to the distant periphery. The revitalization of areas along the Contorno Avenue, including Gamboa de Baixo, another poor black neighborhood in the city center, constitutes a central objective of subsequent stages of Salvador's urban-renewal program. One of Salvador's coastal communities situated on the shores of the Bay of All Saints, Gamboa de Baixo is located on the land underneath the Contorno Avenue. As in the case of the Pelourinho, the people of Gamboa de Baixo do not fit into the government's plans for the area. It intends to remove the existing population in order to implement cultural tourism projects, leisure sites, and new real-estate investments. The construction of the massive luxury high-rise O Morada dos Cardeais in 2005 represents one recent example of the rapid real-estate development of the coastal lands for wealthy residents. In fact, very few traditional poor black fishing communities continue to exist along the bay.

Although the public image of Gamboa de Baixo is that of danger, misery and marginality, as in the case of the Pelourinho, the locale existing below the Contorno Avenue provides the ideal site for the restoration of new urban spaces in Salvador. The government of Bahia began to pay special attention to the neighborhood chiefly because this predominantly black and working-class community occupies some of the most valuable land in the center of Salvador. However, before there was an urban revitalization program for the area, there was a Gamboa de Baixo where people lived and worked for many generations. Formerly known as the Gamboa Port or Porto das Vacas, Gamboa is a fishing colony residents claim to have begun in Salvador's early colonial history as a *quilombo*, a community formed by escaped slaves and indigenous peoples. Before there was the Bahian Metropolitan Development Project, there were more than 350 poor families living in this "coveted" area in the city of Salvador. Though sometimes labeled a recent land "invasion," this fishing colony, for more than a century, has had as its visual reference the Bay of All Saints and, in the distance, the island Itaparica.

A crucial moment in Salvador's modernizing history was the construction of the Contorno Avenue, designed in 1952 and first opened to transit in 1961. It has become one of Salvador's most important roadways. The Contorno Avenue connects the commercial zone of Cidade Baixa (Lower City) and the affluent neighborhoods of Canela in Cidade Alta (Upper City) (Fernandes et al. 2000). At the time of its construction, newspapers reported that the new road constituted one of the most beautiful of all Bahian urban streets because of its picturesque view of the Bay of All Saints, which defines the city of Salvador. The Contorno Avenue represented Bahia's most "modern dimension" and was the ideal model for basic roadways in Salvador (Fernandes et al. 2000: 353). Its construction was the example of two essential processes of achieving the goals of urbanization in Salvador: "the technical, displayed in the search for fluidity and healthfulness of physical and social environments," and "the aesthetic, instrumentalized in a perspective of formation of a new city and a new sociability" (Fernandes et al. 2000: 167). A journalist for the local newspaper *A Tarde* wrote in 1959, during the road's construction, "This avenue, besides serving to decongest traffic and stopping Salvador from being just a 'city of only one road', will be more than a tourist attraction to contribute to our capital, endowed with so many natural resources waiting for enjoyment and valorization." As a technical and aesthetic improvement to the capital city, the Contorno Avenue expanded the commercial center as well as the social and environmental aspects necessary to increase tourism in a developing urban Salvador.

Most authors wrote about the Contorno Avenue in terms of natural and technical beauty, progress and economic prosperity. Few mentioned the impact its construction had on the black population that lived in this coastal region. As older Gamboa de Baixo residents remember, the government did not consider them in either the planning or the construction phases of the Contorno Avenue. To complete the project, development engineers demolished various homes, and the government undercompensated owners for the huge amounts of land appropriated. The construction of the road laid the groundwork for further problems communities located along the Contorno Avenue faced in the 1990s with urbanization programs. A reporter for *A Tarde* mentioned briefly in 1961 that near the road's opening, "notwithstanding a house stands intact in the middle of the street, though in precarious situation due to the demolitions of the neighboring houses." The article also reported that the engineer directing the construction project informed government officials that "the proprietor asked for a very elevated price for compensation."

In 1969, *A Tarde* presented a disparaging report that described the population living literally "below the street" after the completion of the Contorno Avenue. The journalist depicted the local community below the street as primarily comprised of thieves, prostitutes, and the "feeble-minded" who lived in dilapidated and overcrowded housing structures without basic sewer systems, running water or electricity. The more dangerous, he claimed, lived in or near the street's concrete arcs. In contrast, he wrote of hundreds of other families (including civil servants and students) who lived on small streets closer to the beachfront and in the ruins of the old navy fort. Though his piece displayed a similar dismay with the lack of "sanitary conditions," he did recognize that in Gamboa de Baixo there existed a small fishing settlement that participated in an active fish-producing economy in Salvador. In general, the article expressed disapproval of what he called a "*favela* of marginality" existing in the plain center of a reviving city. Gamboa de Baixo's social and geographic location, he communicated clearly, was a "challenge to a civilized city." The journalist's portrayal is only one example of misinformation and negative stereotypes about Gamboa de Baixo the print media has generated since the days of the Contorno's construction. Disparaging public opinions have lingered from those times. This form of sociospatial exclusion is the basis of urban renewal programs attempting to revitalize this coastal region and exclude traditional residents (Espinheira 2008).

As the newspaper reports suggest, on the surface, "decayed" urban communities such as Gamboa de Baixo remind the rest of Salvador of its undesirable past. Pejorative descriptions of Gamboa de Baixo as an uncivilized element of the city holding back a thriving and continually urbanizing Bahia have had material consequences for local residents. "Gamboa was one community before the construction of the Contorno Avenue in the 60s and 70s," one elderly woman who has lived there since her birth explained. Gamboa "was marginalized by the passage of the Contorno Avenue, bringing to the residents oblivion and even discrimination in relationship to the rest of the city," residents wrote in a 1996 communiqué. The Contorno Avenue divided the Gamboa neighborhood as Gamboa de Baixo (Lower) and Gamboa de Cima (Upper), a spatial separation that has upheld hierarchies of racial, social, and economic differences between the two neighborhoods. The construction of the street constrained the previously unrestricted movement of ideas, labor, and goods. Only in the mid-1990s did the government construct a staircase that provided access from Gamboa de Baixo to the rest of the city. Most residents can still remember the difficulty of climbing the shaky wooden stairs to reach the Contorno Avenue. Infrastructural changes within the city separated and isolated them as "those below (de baixo)." Elena, a woman in her mid-thirties who has always lived "below the street," pointed out that some women work as domestics in Gamboa de Cima and nearby neighborhoods such as Vítoria that literally look down on their homes in Gamboa de Baixo. Even today, identifying Gamboa de Baixo as your place of residence might prevent you from getting a job, because, as she also claims, some employers "still think we're all thieves." Sentiments of inferiority and superiority have run deep since the Contorno's separation of Gamboa de Baixo and Gamboa de Cima, demonstrating that the project had more than symbolic significance. In stark difference from the "flourishing" city center above the street, the public relegated the community "below" to a cluster of undesirables who lingered behind in both space and time.

In the decades following the construction of the Contorno Avenue, the threat of displacement and dislocation has plagued Gamboa de Baixo. Less than a five-minute walk from Gamboa de Baixo along the street, the Bahia Marina yacht club and the Museum of Modern Art Park of Sculptures replaced the local black community of Preguiça in 1995. Today, visitors can sit on benches in the waterfront park admiring displays of Bahian "modernity" in the form of abstract art without any remnants of the more than seventy-five families that were relocated to a neighborhood in the periphery. In addition to the restoration of the São Paulo Fort in Gamboa de Baixo, the revitalization plans of the Contorno Avenue includes the installation of stores selling nautical products and a park with restaurants, bars and kiosks inside the arcs of the street where visitors will be able to enjoy the view and the seafood of the Bay of All Saints. However, the government does not envision the presence of the people of Gamboa de Baixo in the new environment (as in the case of the Pelourinho and Preguiça), but, rather, it intends to remove the existing population.

THIS STRUGGLE BELONGS TO ALL OF US (DE TODOS NÓS)!

The exclusionary process of modernist urban revival in Salvador has forged grassroots solidarity and mass political mobilization in black communities. As I described in the opening narrative, direct action protest such as the closing of the Contorno Avenue is the most effective tactic of political struggle for Gamboa de Baixo community activists in their movement against land expulsion and for neighborhood improvement. At public meetings and street protests, black women have gone directly for the "real clash," thereby engaging in confrontational politics. Another such example of confrontational politics occurred in August 2004 when residents ambushed the entrance of the state water company EMBASA in the Federação neighborhood (Perry 2009). A group of women led the surprise protest armed with whistles, banners, and a megaphone to vocalize their human right to clean water and adequate basic sanitation in their neighborhood. In direct confrontation with these city officials, community activists forced the local government to guarantee poor blacks' access to vital resources such as potable water.

The Gamboa de Baixo protests transform the ways in which we conceptualize black mobilization and resistance, particularly our understanding of black antiracism struggles in Brazil. "Getting things done" for poor black women in Gamboa de Baixo has meant that, when necessary, they must collectively "get in the face" of the powerful and demystify their power and control. This political approach is unlike the *culturalist* tendencies that Michael Hanchard argued more than a decade ago are the definitive characteristics of black activism (1994: 21, 139, my emphasis). Hanchard observes that where black activists focus exclusively on the politics of Afro-Brazilian cultural practices (Candomblé, Samba, Feijoada), they have been unable to organize a mass political movement aimed at transforming institutionalized forms of racial inequality. He also writes that there exists "no Afro-Brazilian versions of boycotting, sit-ins, civil disobedience, and armed struggle in its stead" (1994: 139). However, Gamboa de Baixo protests illustrate that black activists *do* engage in acts of civil disobedience and violent struggle. Black women's leadership in Gamboa de Baixo's political organization reveals the ways in which they use race and gender to mobilize their community. What is clear is that black women's experiences with marginality and exclusion are one source of political mobilization (Caldwell 2007; Goldstein 2003; Hautzinger 2007; McCallum 2007; Sheriff 2001). Blacks in Salvador make use of their awareness of racism and fight to transform racial and class hierarchies.

The Gamboa de Baixo neighborhood association, Associação Amigos de Gegê dos Moradores da Gamboa, was founded on October 7, 1992. Black women started the organization to establish collective governance and representation for the community in their demands for vital resources such as clean running water and improved sewer systems. The fear of possible expulsion from the coastal area that I described in the previous section was a key factor in pushing forward the grassroots struggle against spatial expulsion. More importantly, it was because "there were women, a dozen or so women, that began to cause alarm, to shout 'look what's happening' " (Alicia, personal interview 2000). The participation of women in the community-based struggle has always been greater than that of men. Indeed, women were often the *only* organizers. As Maria claims, "our association fought without fear," and when responding to the typical questions about the relative absence of men in their organization states that "it is our women who are going to fight and achieve these greater objectives" for the entire community. Women claim that this is because they were more conscious of the short- and long-term impact that land expulsion and relocation to the periphery would have on their families. The centrality of women's participation and their gender consciousness fuel the community movement. Former president of the neighborhood association, Ana Cristina, states that

> It is as if the women were defending their territory . . . Why? It's like I was saying, Gamboa has its own culture, its way of life. Gamboa is one family, and we know, and we think, leaving Gamboa to go somewhere else means being in another environment with another family, being in a place where we don't, people don't know each other, then, new relationships . . . I think that the women got to see Gamboa and the environment is a way of surviving. It is the natural environment of Gamboeiros. We need this here. (Ana Cristina, personal interview 2000)

Like Ana Cristina, several of the women I spoke with in Gamboa de Baixo associate their political awareness in this situation with the recognition of their own differential knowledge as women, and for some, as mothers. From this perspective, *this thing* that empowers black women in grassroots movements around issues of survival is exactly what they know about life and their position in the world. For instance, what women in Gamboa de Baixo describe as a broader preoccupation with "their territory" is a complex understanding of the depth of everyday social and economic conditions that define their existence in a poor neighborhood in the center of Salvador. As stated above, they claim territorial rights and "know" the essentials of living and surviving in difficult conditions, including disease and land expulsion. The *thing* that black women have is the will and wisdom to survive the racially determined socioeconomic inequality of urban spaces.

Consequently, the centrality of black women's participation in this grassroots struggle promotes the articulation of racial knowledge, consciousness and resistance. The political organization brings to the attention of Gamboa de Baixo the practices of institutional racism in the processes of land expulsion. In addition to working with other local neighborhood associations fighting against urban removal, Gamboa de Baixo's neighborhood association finds political support from NGOs and race-based organizations such as United Black Movement and the Black Union for Racial Equality. Although revitalization programs are never discussed publicly by the state in terms of being racial projects (Pinheiro 2002; Pinho 1999), black women confirm their racial and class claims when they link their struggle with other targeted black communities, identifying blackness as a pattern in urbanization programs. Through working with other communities around similar issues of social and economic rights, women acquire a broader consciousness of shared experiences with racial injustice as black and poor people.

> Residents, we need to stay mobilized and alert for the violent and arbitrary actions that are being taken by the mayor and the state government . . . When they announced the cleansing, before the elections, it was not just trash that they want to remove from the center of the city, but also the blacks, the poor people, the beggars, the street vendors, the street children and everything that they think dirties the city. We are not going to let them treat us like trash. We are working people and we have rights. (Gamboa de Baixo community bulletin 1997)

Racial and gender solidarity is a crucial approach to strengthening their community movement. In Salvador, women activists in Gamboa, working in alliances with similar organizations, speak out against the "cleansing" of poor blacks from the center of the city and associate "slum clearance" with "black clearance." Their struggle reflects the racialized circumstances in which they organize and the racialized conditions against which they fight. Gamboa de Baixo is an example of black Brazilians' recognition, as blacks, that they bear the impact of urban revitalization. Black women realize that they are "capable of changing history" and voice concerns against the clearance of urban land. They organize politically against the displacement of thousands of poor blacks to the periphery of Salvador that worsens their already difficult economic situations. During these moments, when even city officials have said in public forums "they didn't think that 'those black women were going to speak' " (Dona Ladi, personal interview 2000), the black women of Gamboa de Baixo actively speak out against "slum-clearance" programs targeting black communities. Black women's participation in this social movement is an important assertion of their voice in urban-space discourses that previously had silenced them.

For blacks, contesting racial domination has meant reclaiming collective power through redefinitions of blackness. Reconstructing political identities based on their own understanding of themselves as black is a source of black women's empowerment necessary for political action. As Dona Nice explains,

> I thought it was really important to speak about our pride in our skin, in our color, in our race. What I liked more was to look in his [any city official] eyes, and say it like this, that 'I am black, with pride'. We didn't go to beg them for anything. We wanted our rights. It's important for us to arrive there and say, I am black, but I am black with pride. I am proud of who I am. I didn't come here to beg from you. I want my rights. The rights are mine. (personal interview 2000)

To be taken seriously as poor blacks is an important task for the Gamboa de Baixo community and political organization. Women often explain that their participation in this movement transforms their previous sense of powerlessness as poor black women within the racist structures of urban governance. Dona Nice also mentions that

> If they slammed their hands on the table, we slammed loudly too, looking at them in their faces, things I would not have done before and today I do them . . . I learned that we can't hold our heads down because we're poor, because we're black women. (personal interview 2000)

Despite their experiences with disrespectful treatment in their interactions with city officials and police violence, these women find power in the public assertion of their racial and gender identities. In consideration of black women's position at the absolute bottom of the social strata, their actions during meetings and protests mark the struggle to counter their everyday indignant experiences with racism and sexism in the public sphere. More significantly, black women's actions "in the face" of the government seek to bring attention to the racist core of urban displacement and resistance.

Asserting political power and reconstructing the image of this black community in the center of Salvador is one important aspect of the social movement for permanence in the area. Gamboa de Baixo's resistance is primarily in defense of citizenship rights for use and control of urban land by the inhabitants. However, Gamboa de Baixo's political organization has fought to prove their legal ownership of the land. Only a few residents have documentation of their ownership. Claiming "native" rights to the land, they rejected official discourses that the community is merely an "invasion" of marginality or an illegal "squatter settlement" estimated to be less than thirty years old. A Gamboa de Baixo activist, Ivana, contests this term as inaccurate, considering the history of the population.

> I do not see Gamboa as an "invasion" . . . There are already six, seven generations of the same family in Gamboa. That is a lot of time to say that it was invaded, that we invaded. Logically, there were other people who came from the outside, but those are few . . . the great majority live here since their grandfathers, grandmothers came here or their grandparents were born here . . . That's why when they say it is an invasion, I

> fight with them [eu brigo] . . . Of course, there weren't this many people here, but it is not an invasion. (personal communication 2000)

Emphasizing an extensive history of residence on the land, Ivana defends their cultural difference and historical particularities. The residents involved in political mobilization against urbanization projects consider families historically rooted on the land, a land they themselves have developed. They have fought to show that Gamboa de Baixo has its own culture that has developed on the coast and in the fort.

Dona Detinha (aged 78) recounts in a recent conversation that she came to Gamboa at age ten to live with her father, an officer in the Bahian navy, which gave him a house on the land to live in. She stresses the changes that the community has undergone with the expansion of the population, including the fact that her teenage granddaughters are "so fresh these days" unlike when she was a child. She also defends wholeheartedly the similarity between Gamboa de Baixo and other poor black neighborhoods in the city of Salvador where the police "comes down every minute [*desce toda hora*]." Throughout Bahian history, blacks formed several other fishing communities like Gamboa de Baixo along the coast of the Baía de Todos os Santos. The most notable difference is the presence of the fort—the area of Gamboa de Baixo in which her family still lives. Whereas they had permission to live in the fort area for several decades, her family faces the possibility of removal if the government restores the area. A previously disinterested government that "abandoned" the fort and the land that "they've taken care of" showed interest when they perceived the profitability of the site for Bahian cultural tourism (Dona Detinha, personal interview 2000).

Using history to claim land rights is a common political strategy for urban black communities of resistance such as Gamboa de Baixo. Collective memory of ownership is a useful way for black Brazilians to contest racial hegemony, to use history as an interpretive tool of collective defiance, empowerment, and solidarity (Hanchard 1994: 150–153). Michael Hanchard writes that "other" memories must compete with a "public past" that is itself the result of the ability of a dominant social group to preserve certain recollections, *deemphasize* or otherwise *exclude* others (1994: 151, author's emphasis). History that has been cultivated from a position of marginality operates in opposition to singular notions of history as cultural dominance. Activists such as black women in Gamboa have produced their own collective memory to question constructions of local and national memories articulated by those in power. Yet, as Hanchard argues, whereas these memories are necessary to critique dominant constructions of the past, they are insufficient to "overrule" contemporary practices of discrimination.

Whereas I recognize some of the limitations Hanchard describes, collective memory *is* the principal means of defining Gamboa de Baixo's identity in relation to this urban space. Historical knowledge functions as an alternative mythmaking process that rearticulates the experiences of subalterns. This approach in black urban communities such as Gamboa de Baixo is a significant form of social activism in contemporary Brazil. In fact, they use social memory, a necessary basis of counterhegemony, not just to further culturalist politics but as a basis for engagement in "real clash" politics. In particular, a radical revision of Brazilian local and national history is crucial for Afro-Brazilian collective racial claims to land in urban communities.

Making historical claims, Gamboa de Baixo activists demand the legalization of individual and collective land ownership and the authorization of its permanent use and control. They also demand the cancellation of revitalization programs for the coastal areas that involve the removal of the local inhabitants. This demand of land ownership considers urbanization a necessary part of permanence, but urbanization defined as their "citizenship right to better conditions for survival" (Selena, personal interview 2000). The Gamboa de Baixo neighborhood association offers reinterpretations of urbanization projects in the following terms:

> The improvement of the quality of the urban environment and the quality of the lives of the low-income populations of big cities, have as the premise the participation of communities . . . the interests of the communities in question (neighborhood association communiqué 1997, my translation).

Residents redefine urbanization as the "greater integration with other neighborhoods of the city" and not "slum clearance" or black land expulsion. Urbanization, as the Gamboa de Baixo political organization defines it, requires that black and poor people participate as actors in the development of the material conditions of their communities. Decision making includes issues such as the expansion of housing, sanitation, electricity and water. Urbanization in Gamboa de Baixo is not about aesthetic changes for future tourism but for black residents who envision healthier futures on the land. Changes in social conditions are for everyone, and not just for a few. Providing alternative proposals for urbanization, Gamboa de Baixo attempts to transform the view of how Brazilian society works in ways that positively transform poor black communities.

The struggle in Gamboa de Baixo is effective in delaying plans of relocation to the periphery and for the improving of material conditions and resources within the community. Black women activists understand these resources as elements of citizenship rights for black communities. The struggle to gain access to additional resources such as employment has been unsuccessful, and the government has failed to issue documentation of legal landownership. However, activists understand their political efforts to be part of a continued struggle for greater participation in the urban public sphere. Today, there is relative political tranquility. Over the past several years, their forms of social activism have changed as the level of threat and fear of relocation also changes from time to time. They organize fewer protests outside of the community and meet within the community primarily around issues of improvement in infrastructure. Neighborhood activists work on these improvements to assure that residents will continue to belong both historically and presently.

Gamboa de Baixo continues to face the possibility of spatial displacement even with recent state investment in their everyday living conditions such as the construction of hundreds of new homes. As Alicia says about their grassroots movement, "the truth is that the struggle never ended. It's just a little calm right now, but we'll have to start all over again" as the government follows through with plans for waterfront development (personal interview 2000). Another activist, Maria, states that future mobilization in the city will be even more complex and painful than past actions. By complicated, she means,

> Although you live for thirty years in a wooden shack and you love that shack, people have this thing, this sentiment for the material good . . . then, the government comes and says, "no, I give you a brick house in another place," and you leave, they call them "shanties," "leave these 'shanties' and living with rats." And when you are in a brick house, again the government says it will give you a better one in another place. You say, "no, but mine is also made of stone and I like it here, here is where I want to stay." Then, if before you had feelings of love for that place, that house, now, that multiplied, doubled, you know. Because you live in place that you like and that has a certain comfort that works for you, you know, to live, to live well. Then, it's going to be more complex. A lot more. The struggle is going to be more *assirrada* [arduous], much more difficult, and I hope that we are well prepared because that nautical [development] project makes promises, and it promises a lot. The government does not plan to give up because of Gamboa. And Gamboa does not plan to give up because of the nautical project (personal interview 2000).

Activists continue to participate in solidarity with other communities around issues of urbanization and land rights. Everyday struggles revolve around efforts to maintain and increase improved conditions in the community. Nevertheless, as Alicia states above, the future development of Salvador's urban waterfront is still probable, and urban renewal for tourism promises the exclusion of Gamboa de Baixo residents. She predicts that their exclusion will again produce fear and stir powerful sentiments of mass mobilization in defense of their territorial rights. The black struggle for collective permanence and land rights is ongoing.

CONCLUSION: SPATIAL DEFIANCE

Urban spaces are terrains of constant struggle for black and poor people (Gregory 1998; Holston 2008; Kowarick 1994). The Gamboa de Baixo cultural tourism project and the exclusion of its black population demonstrate the discursive and material realities of systemic racial inequality. Official governmental accounts assert that some measure of success has been achieved in the implementation of urban economic and social reform. However, an alternative assessment leads to the conclusion that urban renewal often has been far less than a panacea to economic and social ills. As multinational investors fund urban-renewal projects designed to alleviate "visible" poverty and facilitate rapid economic growth around cultural tourism, the government has increasingly expelled "*o negro pobre do centro da cidade* [the black poor from the center of the city]" to its margins.

Specifically, I illustrated that urban development projects are oftentimes institutionalized racial projects of spatial exclusion. The most worrisome aspect of Salvador's ongoing urban redevelopment is the violent disappearance of visible black clusters in the center of this vibrant city. The underlying logic of urban development in Brazilian cities is racial exclusion. As Thomas Sugrue (1996: 229) writes, urban space is a "metaphor for perceived racial difference." The strategies of redevelopment operate as institutional mechanisms to maintain *de facto* racial and class segregation (Dávila 2003; Davis 2007; Jackson 2003; Pearlman 2010). The case of Gamboa de Baixo illustrates the ways in which the conceptualization and implementation of these renewal and revitalization programs in Salvador exclude and displace black communities socially, economically, and spatially. City redevelopment programs exclude blacks and relocate them to the distant periphery of new urban spaces. Growing gentrification "transforms" cities, but also deepens racial and class divisions by spatially demarcating the socioeconomic boundaries of racially ordered spaces. This connection between exclusive development and racial order makes explicit the racism propelling this model of exclusion. For Salvador's leaders, aesthetic and economic development depends on the exclusion of blacks and their subsequent relocation to the city's geographical periphery. Paradoxically, the presence of blacks is not just an obstacle to urban modernization. Their presence is necessary insofar as they contribute to the reification of Brazilian national identity through commodified minstrelization, folklorization, and fetishization of black culture and history. However, without inclusion of the communities which traditionally produce that culture, there is no space for black people themselves (Vargas 2008).

Political movements have emerged in response to unequal socioeconomic and racial segregation in city planning. Gamboa de Baixo's political organization is just one example of a black community actively engaged in protest against the racist politics of exclusion underlying urban revitalization programs. Organizing as blacks, they have led the attack against "slum clearance" or the "cleansing" of blacks and poor people from economically profitable spaces. This ethnographic analysis of the grassroots organization in Gamboa de Baixo illustrates how blacks *have* mobilized politically on the basis of black identity and in pursuit of concrete political objectives centered on their rights to land. In the Bahian urban center, permanent territorial rights constitute a local idiom for the affirmation of black consciousness and cultural insurgency. Contrary to Michael Hanchard's analyses of black politics (see also Butler 1998; Caldwell 2007; Covin 2006; French 2009), the reformulation and reinterpretation of black cultural identities in contemporary northeastern Brazil *do* translate into social and economic projects that resist institutional racism embedded in projects of urbanization. Black grassroots organizations such as the neighborhood movement of Gamboa de Baixo have identified racism as an aspect of social inequality in Brazilian urban communities. The fact that blacks feel that they are disproportionately subjected to urban removal and land expulsion is a central issue in black political organization. These urban communities conceptualize their experiences with marginalization as a facet of black racialization. As a result, blacks use their racial, gender, and class identities to protest against institutional mechanisms of racial inequality that utilize spatial displacement as an acceptable form of urban transformation.

In sum, urban revitalization programs in Salvador, Bahia, Brazil are racist social policies against poor blacks. I showed the ways in which the centrality of black women's participation in grassroots organizing has influenced the articulation of racial and gender-identity politics in urban struggles against spatial displacement and relocation. Focusing on activists' narratives and actions of contestation, I establish the idea that black social activism in Brazil has emerged, primarily led by women at the community level in struggles for access to material resources in urban spaces such as land and housing. Activists explain that they demand the resources they need to live and take care of their families that are available to them in the city center. In Gamboa de Baixo, a major question is "What will we do someplace else, someplace we do not know?" which illustrates residents' keen understanding of their spatial location in relationship to how displacement from the city center further diminishes their ability to "survive." From this viewpoint, grassroots activism in Salvador's black neighborhoods forces us to rethink black resistance as well as to reconsider the ways in which blacks offer alternative views on the way Brazilian society operates and should operate.

NOTE

1. This is a revised and updated version of my previously published essay (2004), "The Roots of Black Resistance: Race, Gender and the Struggle for Urban Land Rights in Salvador, Bahia, Brazil." *Social Identities* 10 (6): 7–38.

REFERENCES

Butler, Kim D. (1998). *Freedoms Given, Freedoms Won: Afro-Brazilians in Post-Abolition São Paulo and Salvador.* New Brunswick, NJ: Rutgers University Press.

Caldwell, Kia Lilly (2007). *Negras in Brazil: Re-Envisioning Black Women, Citizenship, and the Politics of Identity*. New Brunswick, NJ: Rutgers University Press.

Collins, John (2008). "But What If I Should Need to Defecate in Your Neighborhood, Madame? Empire, Redemption, and the 'Tradition of Oppression' in a Brazilian World Heritage Site." *Cultural Anthropology* 23 (2): 279–328.

Covin, David (2006). *Unified Black Movement in Brazil, 1978–2002*. Jefferson, NC: McFarland and Company.

Dávila, Arlene (2003). *Barrio Dreams: Puerto Ricans, Latinos, and the Neoliberal City*. Berkeley, CA: University of California Press.

Davis, Mike (2007). *The Planet of Slums*. New York: Verso.

Dunn, Christopher (1994). "A Fresh Breeze Blows in Bahia." *Americas* 46 3: 28.

Espinheira, Gey, ed. (2008). *Sociedade Do Medo*. Salvador: EDUFBA.

Fagence, Michael (1995). "City Waterfront Redevelopment for Leisure, Recreation, and Tourism: Some Common Themes." Pp. 135–155 in *Recreation and Tourism as a Catalyst for Urban Redevelopment: An International Survey*, edited by Stephen J. Craig-Smith and Michael Fagence. Westport, CO: Praeger.

Fernandes, Ana, Sampaio, A.H., Gomes, M.A.A.F. (2000). "A Construção do Urbanismo Moderno na Bahia, 1900–1950: Construção Institucional, Formação Profissional e Realizações," Pp. 167–182 in *Urbanismo no Brasil, 1895–1965*, Maria Cristina da Silva Leme, Ed.: São Paulo: Livraria Nobel, S.A.

French, Jan Hoffman (2009). *Legalizing Identities: Becoming Black or Indian in Brazil's Northeast*. Chapel Hill: University of North Carolina Press.

Goldstein, Donna M. (2003). *Laughter Out of Place: Race, Class, Violence, and Sexuality in a Rio Shantytown*. Berkeley: University of California Press.

Gregory, Steven (1998). *Black Corona: Race and the Politics of Place in an Urban Community*. Princeton, NJ: Princeton University Press.

Hanchard, Michael (1994). *Orpheus and Power: The Movimento Negro of Rio De Janeiro and São Paulo*. Princeton, NJ: Princeton University Press.

Hautzinger, Sarah J. (2007). *Violence in the City of Women: Police and Batterers in Bahia, Brazil*. Berkeley: University of California Press.

Holston, James (2008). *Insurgent Citizenship: Disjunctions of Democracy and Modernity in Brazil*. In-Formation Series. Princeton, NJ: Princeton University Press.

Jackson Jr., John (2003). *Harlemworld: Doing Race and Class in Contemporary Black America*. Chicago: University of Chicago Press.

Lefebvre, Henri, Eleonore Kofman, and Elizabeth Lebas (1996). *Writings on Cities*. Cambridge, MA: Blackwell Publishers.

McCallum, Cecilia (2007). "Women out of Place? A Micro-Historical Perspective on the Black Feminist Movement in Salvador Da Bahia, Brazil." *Journal of Latin American Studies* 39: 55–80.

Marcuse, Peter, and Ronald Van Kempen (2000). *Globalizing Cities: A New Spatial Order?* Malden, MA: Blackwell Publishers.

Nascimento, Abdias do (1989). *Brazil: Mixture or Massacre? Essays in the Genocide of a Black People*. Dover, MA: First Majority Press.

Lacarrieu, Monica (2000). *"No Caminho Para O Futuro, a Meta E O Passado". A Questão Do Patrimônio E Das Identidades Nas Cidades Contemporâneas*. V Congresso International da BRASA—Brasil 500 Anos.

Perlman, Janice E. (2010). *Favela: Four Decades of Living on the Edge in Rio De Janeiro*. Oxford and New York: Oxford University Press.

Perry, Keisha-Khan Y. (2009). "If We Didn't Have Water: Black Women's Struggle for Urban Land Rights." *Environmental Justice* 2 (1): 9–13.

Pinheiro, Eloísa Petti (2002). *Europa, França E Bahia*. Salvador: EDUFBA.

Pinho, Osmundo de Araujo (1999). "Espaço, Poder E Relações Racias: O Caso Do Centro Histórico De Salvador." *Afro-Àsia* 21–22: 257–274.

Pinho, Patricia de Santana (2010). *Mama Africa: Reinventing Blackness in Bahia.* Durham, NC: Duke University Press.
Romo, Anadelia A. (2010). *Brazil's Living Museum: Race, Reform, and Tradition in Bahia.* Durham. NC: University of North Carolina Press.
Sansi, Roger (2007). *Fetishes and Monuments: Afro-Brazilian Art and Culture in the Twentieth Century.* Oxford: Berghahn.
Sheriff, Robin E. (2001). *Dreaming Equality: Color, Race, and Racism in Urban Brazil.* New Brunswick, NJ: Rutgers University Press.
Sugrue, Thomas (1996). *The Origins of the Urban Crisis: Race and Inequality in Postwar Detroit.* Princeton, NJ: Princeton University Press.
Vargas, João Costa (2008). Never Meant to Survive: Genocide and Utopia in Black Diaspora Communities. Lanham, MD: Roman and Littlefield Publishers.

Epilogue

Second May Be Best—Theorizing the Global Urban from the Middle

Xiangming Chen and Michael Magdelinskas

Instead of a more conventional conclusion that can echo the introduction in making an edited volume coherent, we have opted to offer a shorter epilogue to reflect on the most important lessons that can be distilled from the preceding chapters and to also bring clarity to their collective contributions. We do so from the premise and rationale for this book: bringing together a set of less studied, secondary (see Map I.1 again) cities and going beyond making them more familiar and important for their own sake, but also making sufficient sense out of them with the intention of contributing to a richer theorizing of global urbanism.

REFRESHING SCALE AND POWER

While the urban literature has become more comparative, diverse, and global over the last two decades or so, it has sustained a strong focus on scale and power, often in combination, as two salient features that draw research attention to certain cities. In one major strand of the literature, the study of world or global studies continues to be shaped, perhaps justifiably so, by the frameworks developed by John Friedmann and Saskia Sassen (Kanna and Chen introduction, to this volume), emphasizing the economic power of a number of world cities (not just New York and London) in a global hierarchy or network. Along a different research trajectory, scholars have brought a number of economically rising cities of the global South that are often, but not exclusively, of megascale, like Shanghai and Mumbai, under the analytical microscope. The two editors of this book are recent contributors to this discourse, with their works on Shanghai and Dubai, respectively (Chen 2009; Kanna 2011). This sustained emphasis on scale and power excludes many cities that may not meet either or both criteria. We end up paying a price for missing the empirical and theoretical insights that can be extracted from studying those excluded cities.

To redress the bias toward large scale and great power in choosing what cities to study, we first need to get a glimpse of the relative weights of different-sized cities in shaping the present and future urban world. According to UN-HABITAT (2008), in 2000, 60.7 percent of the urban population in developing countries lived in either intermediate (500,000–1 million) or big (1–5 million) cities, while large (5–10 million) and mega- (more than 10 million) cities accounted for only 17.7 percent of the total. In addition, 47.2 percent of intermediate cities and 47.7 percent of big cities experienced high growth (2 to 4 percent per annum) or very high growth (above 4 percent per annum) during 1990–2000, compared to 42.1 percent for large and megacities, which also had a larger percentage of them in slow growth or decline. In all United Nations' forecasts on urban growth to 2030 and beyond, intermediate and big cities (with less than 5 million people) will account for the largest share of urban growth in developing countries. In both China and India, which will contribute the most people toward future urban growth in developing countries and the entire world, intermediate or midsized cities (a size designation by China for those cities with populations up to 5 million) will absorb almost half of their urban expansion. The economic importance of these cities "in the middle" will also rise. Midsized Chinese cities will account for 34 percent of the GDP in 2025, up from 29 percent in 2007 (Devan, Negri, and Woetzel 2008). By 2030, of the 1.3 billion more people and consumers who will live in developing-country cities, only about a fifth of them will reside in megacities, whereas almost all the rest will spread out among 1,100 cities with populations over 500,000, up from just over 700 such cities today (*Bloomberg Businessweek* 2010). This demographic force will shift the economic tide toward the middle range of the global urban hierarchy.

Besides their aggregate demographic and economic significance, each of the midsized cities, sometimes also called second-tier or secondary cities, face a similar set of challenges in different ways: sufficient economic growth, creating jobs for large numbers of in-migrants, balancing between wealth and poverty, improving urban governance, social conflict, spatial segregation, and dealing with historical legacy and recent globalization. We simply do not know enough about how this large category of cities and their diverse communities and residents experience and respond to these varied challenges because we have not studied enough of them, and rarely through a comparative lens on the different facets of these issues. This empirical neglect has limited our ability to theorize the wide variations in global urban processes along and beyond the dominant vectors of scale and power. With the set of contributions in this book, we have focused squarely on the large middle of the urban hierarchy—represented by a diverse set of midsized or secondary cities—as cases and contexts for probing several thematic topics. Except for Shenzhen, China, in Chen's and de'Medici's chapter, Tianjin, China, in Liu's and Chen's chapter and New York and Shanghai as contexts in Zukin's chapter, none of the cities, even including their metropolitan regions, covered in this book exceeds five million people. Yet all of them have offered substantial substantive and contextual details regarding the treated topics that have helped us theorize global urbanism beyond scale and power.

LEARNING FROM THE MIDDLE

What can secondary cities, or those in the middle range of the global urban hierarchy, teach us that we would not otherwise learn from and about? First of all, their relative status makes them vertically connected to the first-tier cities on the global hierarchy scale. This conceptual view tends to bias research toward the top of the hierarchy to look down. Resisting to look at the diverse cities in this volume hierarchically (Bunnell and Sidaway, preface for this volume), a few authors have not only placed their chosen secondary or midrange cities horizontally onto the global circuit of ideas and practices that flow and traverse across national boundaries, but also located them deeply into the local cultural and economic milieus. Following the Pompidou Center as a cultural strategy for urban redevelopment in central Paris, the Spanish city of Bilbao (population around 350,000) built a Guggenheim—an outpost of the global chain of the New York–based Guggenheim Museum—to revive an old declining area (Zukin, this volume). Whereas Shanghai has followed the lead of New York and Hong Kong in developing its own creative artistic hubs, second-tier Chinese cities like Chongqing (although not in demographic scale) in the southwest have imitated Shanghai's cultural strategy of urban renewal by building their versions of Shanghai's Xintiandi, which is a glamour district of luxury shops, fancy bars, and high-end restaurants with a traditional local architectural façade built on an old and torn-down central-city site. One could argue there is a top-down spread of urban cultural strategies from New York to Paris to Bilbao and from New York to Shanghai to Chongqing. The horizontal flow of urban governance ideas and practices among secondary cities manifests itself both in Johannesburg, South Africa, and Harare, Zimbabwe, adopting participatory budgeting from Porto Alegre (population 1.5 million), Brazil, albeit with differential success in their respective African local contexts (Masiya, this volume). These studies demonstrate that secondary cities are much more integrated into the global network of flows by design, instead of being passive recipients of trickle-down influence, than has been recognized.

We have learned more about the horizontal learning between secondary cities and their stronger than expected participation in the global economy via a deeper and more updated understanding of historical legacy revealed by a couple of chapters in this book. Through a rigorous and nuanced reading and reaccount of the parallel, but different, histories of Dubai and Singapore, Kanna (this volume) has traced the neoliberal development policies of the two city-states to their similar roots in British colonial presence and influence in both regions and to their postindependence practice of certain colonial governance ideologies and principles. Besides sharing this colonial past and its inertia, Singapore adopted some developmental state policies that originated from Japan and then spread to Korea and Taiwan in the 1970s, whereas Dubai followed a little lead from Singapore in leveraging its position as an entrepôt between Asia and Europe in the 1990s. In a totally different geographic and temporal context, Chinatown politics in San Francisco in the middle of the twentieth century, with its deep divide and conflict between the pro-Communist and pro-Nationalist factions, suggests that ethnic leaders' relations with business and government were both local and transnational in nature (Tang, this volume), long before what the recent diasporic literature has confirmed about the strong association between the economic rise of China and the revival of Chinatowns in a number of American and European cities beyond New York. We have learned here that the urban legacy of economic and political histories weighs heavily on some secondary cities even if their strong global connections today seem to mark a clear breakaway from past trajectories.

The next lesson we have learned picks up, and then extends, from the one above. It pertains to how the study of secondary cities can help us critique the dominant globalization and global city discourse that features hierarchical power and inter-city competition, in light of complicating historical and political forces operating at the subnational and regional levels. Whereas global competition has undoubtedly eroded manufacturing in old American cities of the Northeast and Midwest, local and regional responses to this grave erosion have varied considerably, reflecting the severe limits to, and the big failures of, urban regeneration strategies no matter how radical they may be. Despite their very different scales and industrial traditions, both Springfield, Massachusetts, and Detroit, Michigan, have experienced a protracted decline in machine-tools and automobile manufacturing for the last several decades, respectively (Forrant, this volume; Ryan, this volume). The two cities, however, diverged in the major strategies they have used to deal with the budgetary malaise of lost tax bases. Whereas Springfield in 2004 negotiated with the state government to accept an appointed control board that would make future spending decisions, Detroit, in cross-border cooperation with Windsor, Ontario, invested heavily in building four casinos as an alternative economic engine from the late 1990s into the 2000s. Neither approach has worked well. The Control Board of Springfield has done little to stimulate economic development, whereas much of the casino revenue in Detroit has come from the accumulated savings of local residents and flown into the casino corporations instead of to the local government coffers. As both case studies suggest, Springfield and Detroit have ironically become mired in a *de-globalizing* (Ryan's term) groove, which makes it extremely difficult, if not impossible, for them to revive themselves. In a completely opposite direction, Tianjin Binhai New Area (TBNA) has experienced explosive economic growth by using a state-sponsored *globalizing* strategy featuring financial incentives that have attracted a lot of foreign investment (Liu and Chen, this volume). Due to TBNA's appeal as a newly developing locality in China, its government-provided financial incentives have achieved the intended results, whereas the same approach by many state and local governments in the United States has only led more businesses to do "incentive shopping," especially since the global economic crisis in 2008, without generating much new net investment in local economies.

Whereas three secondary cities here have provided new insights into local economic restructuring under globalizing or de-globalizing dynamics, other such cities covered in this book offer fresh empirical evidence on the spatial construction and manifestation of national identity, social imagination, and political exclusion that challenge our expectations. Whereas the increasingly open and 'borderless' world is supposed to weaken the nation-state, the national governments of Malaysia and Indonesia have devised new strategies to maintain and materialize the 'imagined community' of the nation by designing Muslim-style architecture for their new small cities of Putrajaya and Dompak, respectively (Moser, this volume). Having found taxi drivers in Yunnan, China, perceiving dangerous people or those they don't want to carry as from outside the city, the peasant, or ethnic minority groups, Beth Notar (this volume) old urban-rural divide, a category that should be in retreat under the more open and fluid conditions of China's economic reform. Extending this imagination of space approach to studying patronage at a famed coffeehouse in Tunis, Rodney Collins (this volume) has mapped out segmented sociospatial structures of differences based on how status, class, sexuality, and gender manifested themselves in small spaces. Staying with gender and bringing in race, Keisha-Khan Perry (this volume) has revealed the racist politics of spatial exclusion targeting blacks that are embedded in urban renewal and revitalization programs in Salvador, Brazil. The organized movement against this urban renewal, Perry argues, represents the struggle of poor black communities led by women to protect their rights to material resources, particularly housing and land. Used as diverse contexts for understanding constructed national identity, individually imagined spatial divides and structures, and community politics of exclusion and protest, the secondary cities in Part III of the book have allowed the contributors to paint a rich picture of fragmented tensions and conflicts in the local urban arena. Although not as culturally heterogeneous as the top-tier global cities, the secondary cities like the ones in this book can reveal just as much diversity and fluidity in their small local spaces of identity and politics, imagined or real.

FINAL IMPLICATIONS AND NEXT STEPS

The lessons we have identified above may or may not be sufficiently robust and representative of all secondary cities for advanced and systematic theorizing of the evolving global urban literature, even though they have been derived from a diverse group of cities that have rarely been scrutinized collectively. This design by itself is not necessarily new and significant. It is the combination and convergence of the thematic findings and insights through the multidisciplinary and interdisciplinary lenses that add fresh contributions to the literature. Even without the large-scale and dominant positions of top-tier cities in the global urban hierarchy, several secondary cities in this book are still strongly plugged into the global flows of cultural strategies, urban practices, and economic activities, but in different ways. More importantly, by turning to local histories and politics instead of privileging the dominant global city perspective, some contributors have successfully situated their cases in an alternative framework that favors looking at de-globalizing and resurgent national and local forces as of greater significance to urban restructuring and practices. Some of these unexpected relations have been uncovered from below in community and individual working and living spaces that were examined by the chapters in Part III of the book. Overall, these tendencies unfolding in some secondary cities that run counter to the globalization and global city discourse would not have been detected through a more exclusive focus on top-tier cities.

An obvious next step would be to bring more of the cities we have studied here into the orbit of comparative urban research. But this is not good enough because there are plenty of secondary cities in the big middle of the global urban hierarchy that could be studied. There are three more useful steps to take. One is to think about how to position or pivot secondary cities away from the strong but narrower focus of the large-scale and hierarchical power of the global city perspective so they would not be conceptualized as being passively shaped by vertical structural forces. An effort could also be made to carefully select the secondary cities for study that are more likely to offer either refreshing or challenging evidence on the major debatable or controversial research questions, such as if and how culture, including amenities, matter to urban growth. This is connected to another critical step for research on secondary cities to be based on an intimate knowledge of the local cultural and linguistic contexts and grounded fieldwork (see Bunnell and Sidaway's preface, for this volume). Most of the chapters in this book reflect a rich symbiotic relationship between the researchers and the cities they have studied closely. These suggested steps point to the continuing research direction and agenda that this book has intended to set out.

REFERENCES

Bloomberg Businessweek (2010). "The Untold Wealth of Unknown Cities." October 4–10: 9–10.

Chen, Xiangming, ed. (2009). *Shanghai Rising: State Power and Local Transformations in a Global Megacity*. Minneapolis: University of Minnesota Press.

Devan, Jinamitra, Stefano Negri, and Jonathan R. Woetzel (2008). "Meeting the Challenges of China's Growing Cities." *McKinsey Quarterly* 3: 107–116.

Kanna, Ahmed (2011). *Dubai: The City as Corporation*. Minneapolis: University of Minnesota Press.

UN-HABITAT (2008). *State of the World's Cities 2008/2009*. Sterling, Virginia: Earthscan.

Contributors

Xiangming Chen is Dean and Director of the Center for Urban and Global Studies and Paul Raether Distinguished Professor of Sociology and International Studies at Trinity College in Hartford, Connecticut, as well as Distinguished Guest Professor in the School of Social Development and Public Policy at Fudan University in Shanghai, China. He coauthored *The World of Cities: Places in Comparative and Historical Perspective* (Blackwell Publishers, 2003; Chinese edition, 2005), published *As Borders Bend: Transnational Spaces on the Pacific Rim* (Rowman & Littlefield, 2005), edited and contributed to *Shanghai Rising: State Power and Local Transformations in a Global Megacity* (the University of Minnesota Press, 2009; Chinese edition, 2009), and coauthored *Introduction to Cities: How Place and Space Shape Human Experience* (Wiley-Blackwell, 2012; Chinese edition, 2012). His articles have appeared in such urban studies journals as *City & Community*, *Cities*, *Environment and Planning A*, *International Journal of Urban and Regional Research*, *Urban Affairs Review*, *Urban Geography*, *Urban Studies*, and two dozen edited books.

Ahmed Kanna is Assistant Professor in the School of International Studies at the University of the Pacific. His publications include *Dubai, The City as Corporation* (2011, University of Minnesota Press) and articles in journals such as *Cultural Anthropology*, *City*, *Traditional Dwellings and Settlements Review*, *Review of Middle Eastern Studies*, and *Middle East Report*. He is currently working on the cultural politics and discourses of green development and sustainability in Berlin.

Tim Bunnell is Associate Professor in the Department of Geography at the National University of Singapore. He also holds a joint appointment with the NUS Asia Research Institute, where he leads the 'Asian Urbanisms' research cluster. His most long-standing research interest concerns the politics of urban development in greater Kuala Lumpur, work which included the book *Malaysia, Modernity and the Multimedia Super Corridor* (London: Routledge, 2004). More recently, he has also worked on historical geographies of interconnection between postcolonial Southeast Asia and the city of Liverpool, UK. Tim is currently working on two new projects in Indonesia, one concerned with urban change in an era of decentralization, the other concerning the role of cultural aspirations in shaping city lives.

Rodney W. Collins is a sociocultural anthropologist and research fellow affiliated with the Centre d'études maghrebines à Tunis. His research weaves together ethnographic research with historical analysis to investigate the complex circuitry of market-state-society, especially as it irradiates the formation of contemporary male subjectivities. Collins has recently expanded the scope of his research agenda to include three related domains of inquiry: deaf culture, identity, and the codification of Tunisian Sign Language; youth, gender, and consumption in urban Tunis; and the emergence of the 'one-man show' in the landscape of Tunisian expressive culture. Collectively, these projects work to advance Collins's interest in the critical analysis of social difference at the margins of political and cultural normativity in contemporary North Africa. Collins has been the recipient of Ford, Mellon, Wenner-Gren, and Fulbright grants and has taught secondary school in Cameroon, as well as undergraduate and graduate courses in anthropology, history, and gender and sexuality at Columbia, New York University, and Georgetown.

Tomás de'Medici, a graduate of Trinity College, is pursuing business opportunities expanding client operations throughout China, Latin America, and the United States. While at Trinity, he worked at the Center for Urban and Global Studies. His involvement in the Center lead to traveling the Yangtze River exploring environmental and urban changes and working at the 2010 World Expo in Shanghai with a focus on Sino-Latin relations. He has subsequently been copublished with Dean Xiangming Chen on "The 'Instant City' Coming of Age: The Production of Spaces in China's Shenzhen Special Economic Zone" (*Urban Geography* 2010). Through his travels he has come to speak advanced Mandarin Chinese and conversational Spanish. He resides in the Chicagoland area.

Robert Forrant is Professor of History and teaches in the graduate program in Regional Economic and Social Development at the University of Massachusetts Lowell. Forrant is coeditor of the 'Work, Health and Environment' academic book series for Baywood Publications and is on the editorial board of the journal *Mass Benchmarks*, a joint publication of the University of Massachusetts President's Office and the New England Federal Reserve Bank. His most recent publications are *Metal Fatigue: The Rise and Precipitous Decline of the Connecticut River Valley Industrial Corridor*, Baywood Publishers, 2009; and with Christoph Strobel, *The Big Move: Immigrant Voices from a Mill City*, Loom Press, 2011. Robert_Forrant@uml.edu

Chang Liu is a Trinity senior from Beijing, China, and will graduate in 2012. He majors in economics and plans to pursue a self-designed second major in urban studies. He participated in Trinity's Cities program in his first year and has since worked as a research assistant at the Center for Urban and Global Studies. For his research, Chang has focused on China's urbanization and economic development and produced two publications with Dean Xiangming Chen. He has been awarded a Kenneth S. Grossman Research Grant, a Tanaka Student Research Grant, and is a Long Walk Societies Scholar. In the 2010–11 academic year, Chang studied abroad at the University of Oxford. Chang has been involved in a number of organizations on campus such as the Lion's Club, iHouse, AASA, and the Rising Star program. He will attend law school after graduation.

Michael Magdelinskas received his BA in English and urban and global studies from Trinity College, and is currently pursuing his JD at Cornell Law School. During his time at Trinity, Michael received a number of fellowships and research grants, including the A.V. Davis Summer Research Fellowship and the Kenneth Grossman Global Research Grant. Having studied abroad in Spain, France, and China, Michael hopes to parlay those experiences into his law career, focusing on transnational interactions. A former professional singer, Michael remains involved in the arts and enjoys studying the connections between the arts, urbanization, and the law.

Tyanai Masiya is a doctoral candidate in the Center for African Citizenship and Democracy, School of Government, University of Western Cape. He was a visiting scholar at the Center for Urban and Global Studies at Trinity College during 2008–09. Tyanai has previously taught at Africa University and the Midlands State University, both in Zimbabwe. His previous publications cover local government management in Zimbabwe and politics and governance in Africa.

Sarah Moser holds a PhD in cultural geography from the National University of Singapore. She has held postdoctoral fellowships at in the Aga Khan Program for Islamic Architecture at the Massachusetts Institute of Technology and the Center for Urban and Global Studies at Trinity College. Sarah is currently Assistant Professor of Asian Studies and Culture at the University of Massachusetts Lowell. Her work is broadly concerned with how religious, ethnic and national identities are manifested in the built environment in urban Southeast Asia and her work has been published in such journals as *Social and Cultural Geography*, *Area*, and *Cities: The International Journal of Urban Policy and Planning*.

Beth E. Notar is Associate Professor of Anthropology at Trinity College. She is the author of *Displacing Desire: Travel and Popular Culture in China.* As an anthropologist, her research focuses on the intersection of the cultural and the material. This focus has led her to examine the relationship between representations in popular culture, tourism and transformations of place in southwest China; and money as a symbolic, economic and political object. Her current project examines the rise of automobility in China.

Keisha-Khan Y. Perry is Assistant Professor of Africana Studies at Brown University. Her work specializes in black women's activism, African diaspora studies, critical race and feminist theories, urban geography and politics, and race relations in Latin America and the Caribbean. She has done research in Mexico, Jamaica, Belize, Brazil, Argentina, and the United States. Her most recent work is an ethnographic study of black women's activism in Brazilian cities, specifically an examination of black women's participation and leadership in neighborhood associations, and the reinterpretations of racial and gender identities in urban spaces. She is currently completing a book manuscript entitled *Politics below the Asphalt: Black Women and the Search for Racial Justice* based on that research.

Brent D. Ryan believes that design and planning must work more closely together than ever before to solve pressing urban problems. In particular he is interested in how urban design can improve cities characterized by economic failure, building abandonment, and falling populations, in America and elsewhere. These places, commonly known as 'shrinking cities,' allow us to reimagine urban design beyond conventional market-driven esthetics and practice, and to recover the optimism in both critical design and interventionist state policy that vanished with the end of modernism. Brent's work has been published in numerous urban design and urban planning journals, and his book *Design after Decline: How America Rebuilds Shrinking Cities* will be published by the University of Pennsylvania Press in early 2012. Brent has practiced as an urban planner and designer in New York City, Boston, and Chicago, and has taught at Harvard University and the University of Illinois at Chicago. He holds a PhD in urban design and planning from MIT, an MArch from Columbia University, and a BS from Yale University.

James D. Sidaway is Professor of Political Geography at the National University of Singapore. He previously worked at the University of Amsterdam and at several universities in the UK. In addition to the history and philosophy of geographical thought, his research interests encompass the interfaces between, geopolitics, borders, development and the urban. Recent empirical work on these themes has been conducted in the capital cities of Mozambique and of the United Arab Emirates.

Scott H. Tang is Assistant Professor of American Studies at Trinity College. He is the author of "Becoming the New Objects of Racial Scorn: Racial Politics and Racial Hierarchy in Postwar San Francisco, 1945–1960," in Jeff Roche, ed., *The Political Culture of the New West* (Lawrence, 2008). His current research investigates San Francisco's racial communities during World War II.

Sharon Zukin is Professor of Sociology at Brooklyn College and the Graduate School of the City University of New York. Her book *Naked City: The Death and Life of Authentic Urban Places* (Oxford University Press, 2010) describes how New York neighborhoods and public spaces have gentrified, privatized, and otherwise upscaled during the past thirty years. She is also the author of *Loft Living*, *Landscapes of Power* (winner of the C. Wright Mills Award), *The Cultures of Cities*, and *Point of Purchase: How Shopping Changed American Culture*. In 2007 Sharon Zukin received the Lynd Award for Career Achievement from the Community and Urban Sociology Section of the American Sociological Association.

Index

N

O

For Product Safety Concerns and Information please contact our EU
representative GPSR@taylorandfrancis.com Taylor & Francis Verlag GmbH,
Kaufingerstraße 24, 80331 München, Germany

Printed and bound by CPI Group (UK) Ltd, Croydon, CR0 4YY
01/05/2025
01858577-0001